CHEMISTRY OF INTERFACES

ELLIS HORWOOD SERIES IN CHEMICAL SCIENCE

CHEMISTRY OF INTERFACES

M. J. JAYCOCK, B.Sc., Ph.D.
Senior Lecturer in Surface and Colloid Chemistry
Loughborough University of Technology

and

G. D. PARFITT, B.Sc., Ph.D., D.Sc.
Professor of Chemical Engineering
Carnegie-Mellon University,
Pittsburgh, USA

ELLIS HORWOOD LIMITED
Publishers · Chichester

Halsted Press: a division of
JOHN WILEY & SONS
New York · Chichester · Brisbane · Toronto

First published in 1981 by

ELLIS HORWOOD LIMITED
Market Cross House, Cooper Street, Chichester, West Sussex, PO19 1EB, England

The publisher's colophon is reproduced from James Gillison's drawing of the ancient Market Cross, Chichester.

Distributors:

Australia, New Zealand, South-east Asia:
Jacaranda-Wiley Ltd., Jacaranda Press,
JOHN WILEY & SONS INC.,
G.P.O. Box 859, Brisbane, Queensland 40001, Australia.

Canada:
JOHN WILEY & SONS CANADA LIMITED
22 Worcester Road, Rexdale, Ontario, Canada.

Europe, Africa:
JOHN WILEY & SONS LIMITED
Baffins Lane, Chichester, West Sussex, England.

North and South America and the rest of the world:
Halsted Press: a division of
JOHN WILEY & SONS
605 Third Avenue, New York, N.Y. 10016, U.S.A.

British Library Cataloguing in Publication Data
Jaycock, M. J.
Chemistry of interfaces. —
(Ellis Horwood series in physical chemistry).
1. Surface chemistry   2. Surfaces (Physics)
I. Title   II. Parfitt, Geoffrey Derek
541'.3453   QD506   80-40387
ISBN 0-85312-028-5 (Ellis Horwood Ltd., Publishers — Library Ed.)
ISBN 0-85312-298-9 (Ellis Horwood Ltd., Publishers — Student Ed.)
ISBN 0-470-27013-6 (Halsted Press)

Typeset in Press Roman by Ellis Horwood Ltd.
Printed in Great Britain by Bulter & Tanner Ltd., Frome.

Table of Contents

Preface

Within normal human experience all physical and chemical processes have an interface associated with them, which usually cannot be entirely ignored. Furthermore there are many everyday situations which are very dependent upon interfacial processes: we breathe, for example. In many aspects of the fields of chemical industry, chemical engineering, food science, agriculture and medicine, to name but a few, surface chemical properties and processes are important. Many disciplines are concerned with the study of interfacial phenomena, and too frequently the degree of understanding displayed is too superficial.

This book attempts to describe, as far as possible, those aspects of the fundamentals of the subject that are necessary to the understanding of an interface. It does not try to be encyclopaedic, or to deal with all aspects of all the possible interfaces, but is confined to those areas which are frequently encountered and of which there is reasonable understanding.

After a general introduction to the concepts of interfacial chemistry and the forces involved, there is a description of the application of chemical thermodynamics to the two-dimensional state of matter, indicating the essential differences in the treatment of interphases from those methods normally associated with bulk phases. An analysis of the concept of a 'Gibbs dividing surface' is given because it is fundamental to the understanding of interphases, whilst remaining an abstract notion.

The subsequent chapters are devoted to specific interfaces. The chapter on liquid interfaces is limited to an extent by the comparative lack of understanding of their behaviour on a molecular scale, and in consequence it concentrates on the macroscopic physical properties of liquid interfaces and their measurement. The next chapter is devoted to a description of the solid-gas interface and modelling it on a molecular scale; an approach which is somewhat novel. Finally the chapter on the solid-liquid interface covers both the macroscopic effects which arise when liquid and solid are in contact, and the microscopic events at the interface.

Dr M. J. Jaycock would like to acknowledge the facilities provided by Dr A. Y. Nehru and The Wolfson Institute of Interfacial Technology of The University of Nottingham during a period of sabbatical leave, when his contribution to this book largely was written.

The poor King looked puzzled and unhappy, and struggled with the pencil for some time without saying anything; but Alice was too strong for him and at last he panted out "My dear! I really *must* get a thinner pencil. I can't manage this one a bit: it writes all manner of things that I don't intend —"

Through the Looking Glass and
What Alice Found There

CHAPTER 1

Introduction to interfaces and the forces involved in their formation

1.1 INTRODUCTION AND DEFINITIONS

The basic fundamental concepts of surface chemistry originated in the works of Thomas Young (1805) [1] and Josiah Willard Gibbs (1872) [2]. Young described the 'surface', suggesting that its mechanical properties may be related to those of a hypothetical stretched membrane, this 'membrane' having a tension, the **surface tension**. Its position was chosen such as to make the simple model equivalent mechanically to the complex region that exists between two bulk phases in contact. Gibbs did the thermodynamics, and described a geometrical surface, the **Gibbs dividing surface**, to enable a satisfactory mathematical description to be made of the region. All this happened before molecular dimensions were established, before polymers were recognised, before quantitative measurements of molecular events at surfaces could be carried out, and of course long before the advent of modern spectroscopic techniques with which the chemistry of the surface can be analysed.

We are, in this book, concerned with the boundary region between bulk phases. Where two phases meet is commonly called an **interface** – the term **surface** is used when one of the phases is gas or vapour. But, because the region of the interface has thickness, it is more correct to talk of an **interphase**; this is the region at an interface between two bulk phases over which there is a gradation of property. The term interface is now usually used in thermodynamic arguments where explicit consideration of the boundary region thickness is involved. Several types of interfaces are readily identified: liquid-liquid, liquid-vapour, solid-liquid, and solid-gas. For each there is a free energy change associated with its formation, termed the **excess interfacial free energy** (or **excess surface free energy** for the surface of a solid or liquid in contact with gas or vapour); this represents the excess free energy which the molecules possess by virtue of their being in the interface/surface. The units of surface tension and specific excess surface free energy (specific refers to unit area) are dimensionally equivalent (kg s^{-2}), and the two quantities are numerically equal for pure liquids in equilibrium with their vapour. Similarly, for two pure immiscible liquids which make contact at a plane interface, the excess interfacial free energy equals the interfacial tension. Not so where solids are involved. The specific excess surface free energy of a solid is equivalent to the surface tension of an equilibrium isotropic surface, but in practice the atoms and molecules in a solid surface freshly generated, for example, by cleavage are not in their equilibrium positions, because their mobility is normally very low.

Atomic or molecular forces are manifested at the interface, and much has been done (a) to relate the large variety of measurable parameters (surface tension, heat of immersion, adsorption from solution, contact angle, etc.) to the structure of the atoms and molecules, and (b) to analyse the forces summed across an interface.

Sorption, a term used to denote the separation of a chemical species between bulk and interface, is frequently encountered in surface chemistry. We also differentiate between changes that occur on the external surface and those which involve the interior. Species are **adsorbed** on the outside and **absorbed** on the inside. Take the case of a porous solid — it is rather like a sponge, but the inside structure is very much smaller; many examples occur in nature, and they are also manufactured for specific applications. The pores have dimensions of the order of 1 — 100 nm, and gases are absorbed into them; the concentration of gas in the pore is greater than that in the bulk phase and it is said to be **positively** absorbed. Adsorption occurs on the external surface of the solid, and again the gas is positively adsorbed. **Negative** adsorption is also possible although less common. It occurs in electrolyte solutions when the concentration of ions at the surface is less than that in the bulk solution.

Forces are naturally involved in the sorption process. They may be of the van der Waals type, and the change is termed 'physical', hence **physical adsorption** or sometimes (unfortunately) 'physisorption', to complement the term **chemisorption** (chemical adsorption) when a chemical bond is formed between solid surface and adsorbed molecules. The difference between physical adsorption and chemisorption is usually manifested by the heat of adsorption, which for the former is low (0 — 20 kJ mol^{-1} of adsorbate), and is of the same order of magnitude as the heat of liquefaction. The formation of a physically adsorbed layer is similar to the condensation of vapour to liquid, and such layers, particularly if many molecular diameters thick, behave like two-dimensional liquids. Since this process is related to liquefaction, it only occurs to an appreciable extent at temperatures and pressures near those required for liquefaction.

In contrast, the adsorption heat for chemisorption is high (80 — 400 kJ mol^{-1} of adsorbate), an order of magnitude that reflects the energy of bond formation; chemisorption is complete once a monomolecular layer is formed, and it usually occurs at lower pressures and much higher temperatures than for physical adsorption. It is relatively easy to displace physically adsorbed species from the solid surface by, for example, reducing the pressure, but desorption of those chemisorbed is much more difficult, often requiring high temperatures.

It should be noted that the values of heats of adsorption quoted above are given as a rough guide. For example, the heat of physical adsorption of water on rutile can be as high as 400 kJ mol^{-1}.

We have talked of the boundary 'region' between two bulk phases. It has a thickness. In some cases the adsorption layers are only of the order of the dimension of one molecule − a **monolayer**, and is common at the liquid-vapour and liquid-liquid interfaces, although the phase(s) adjacent to this layer is also perturbed. In other cases several molecular layers are involved − a **multilayer**, such as for a gas adsorbed at relatively high pressure on the surface of a solid. Chemisorption usually takes place in monolayers, whereas physically adsorbed films on solid surfaces can be multimolecular in thickness.

So we have a picture of the interface − not a geometric plane but one having finite thickness (interphase), sometimes a monolayer but more frequently several molecules thick. Across the interface there is an excess free energy which relates to the forces involved in the formation of that interface. Within the interfacial region the atoms or molecules are in constant motion; in many cases they leave and enter the region at high velocities, so that their residence time is very short. For example, in the surface of a liquid a simple calculation [3] shows the average time to be ∼10^{-6} s.

Winter's study, using ^{18}O, of the continuous exchange between oxygen ions in the surface of an oxide and oxygen molecules in the atmosphere, clearly demonstrated that the surface is in constant motion [4]. Pictures of the interface are necessarily static, but we should not forget the dynamics of the situation. The surface of a liquid may be considered to be homogeneous; that is, its properties do not, on average, vary (laterally) across the surface. In contrast, most solid surfaces are heterogeneous, the chemical nature and adsorption characteristics varying from place to place across the surface. Again it is a question of dynamics, and the rate at which atoms or molecules change their relative position.

1.2 FORCES BETWEEN MOLECULES IN BULK MATTER [5,6,7]
The idea that intermolecular energy is the difference of attractive and repulsive terms, and that each varies as the inverse of distance to the powers m and n respectively, is often attributed to Mie [8], who wrote

$$U = \frac{A}{d^n} - \frac{B}{d^m} \qquad (1.1)$$

where d is the distance of separation and A and B are constants. However, it is likely that the idea was fairly well established before that date. For example in a letter [9] from Woodhouse to Clark[†] which was probably written in 1814, such an equation was proposed, and it was appreciated that n must be greater than m.

[†]Clarke was Professor of Mineralogy, and Woodhouse was later Professor of Mathematics at Cambridge.

The best known form of equation (1.1) is that known as the **Lennard-Jones potential** for the interaction between a pair of non-dipolar molecules, such as a pair of inert gas atoms. This has the form

$$U = \frac{A}{d^{12}} - \frac{B}{d^6} \tag{1.2}$$

and is of moderate accuracy. The problem is that the nature of the repulsive term is not well understood, and values of n between 9 and 12 work satisfactorily in many cases. In certain applications the whole repulsive term is replaced by one of the form $A' \exp(-\alpha d)$ where A' and α are constants. This term arises through unfavourable interactions of both electron and nuclear wave functions.

Much more is known about the form of possible attractive terms, where m may be equal to 1, 2, 3, 4 or 6, or a combination of terms, according to the nature of the molecules, atoms or ions interacting. The interaction of two ions corresponds to $m = 1$, that is to Coulomb's Inverse Square Law for the force. The cases $m = 2$ and $m = 3$ correspond to ion-permanent dipole and permanent dipole-permanent dipole interactions, whilst $m = 4$ corresponds to the ion-induced dipole interaction. There are a variety of $m = 6$ terms, corresponding to dipole-induced dipole, and freely rotating dipole interactions, as well as the London dispersion forces between the instantaneous dipoles in atoms or molecules. There are two higher order terms which are of significance. Firstly $m = 8$ for permanent dipole-quadrupole interactions and secondly $m = 10$ for quadrupole-quadrupole interactions.

There are two complicating factors that need to be taken into account. Firstly, all the above discussion refers to the interaction between individual pairs of atoms, ions or molecules. The interaction of macroscopic bodies is the result of the summation of all the interactions. The sum of the pairwise interactions is only a first approximation, and multibody interactions are important, even if no adequate theoretical description is currently available. The second problem is that London dispersion forces exhibit the phenomenon of retardation. The formation of the instantaneous dipole, which gives rise to this interaction, may be considered as associated with electromagnetic radiation of a particular frequency, which is known as the **characteristic frequency**. When the corresponding wavelength is not very much larger than the distance of separation of the atoms or molecules, then the coupling of the instantaneous dipoles with those induced in neighbouring atoms is affected, and the value of m increases to 7. The transition from unretarded to fully retarded forces usually may be considered to occur over a range, say 30-50 nm. As a result of the complexity of matter and these complicating factors, the total interaction energy of two macroscopic bodies, or a molecule with a macroscopic body, no longer follows the simple relationships outlined above.

1.3 FORCES BETWEEN MOLECULES ACROSS INTERFACES

From what has already been said it would be reasonable to assume that the region of change across an interface would only be a few molecular diameters in thickness, beyond which the behaviour reflects that of the bulk material. However, it has been suggested that long range forces exist at interfaces, extending over much larger distances ($\sim$10 – 100 molecular diameters), although the experimental evidence has led to much debate over the years. In our general consideration of the interactions that exist at interfaces we often need only include the London dispersion force (often called the London – van der Waals force) and the retardation effects, together with any electric forces that exist between ions and permanent dipoles in the system.

The surface tension of a liquid γ^{LV} is a direct measure of the intermolecular forces. If only dispersion forces are involved then $\gamma^{LV} = {}^d\gamma^L$. But in many liquids this is not the case. For example, several types of intermolecular forces exist in liquid water (W), all of which contribute to the surface tension γ^{WV}. Fowkes [10] proposed that we write

$$\gamma^{WV} = {}^d\gamma^W + {}^h\gamma^W \tag{1.3}$$

where ${}^d\gamma^W$ is the contribution from dispersion forces, and ${}^h\gamma^W$ that due to hydrogen bonding and dipole interactions. For mercury (M) there is a contribution due to metallic bonds, hence

$$\gamma^{MV} = {}^d\gamma^M + {}^m\gamma^M \quad . \tag{1.4}$$

The interface between two immiscible liquids, for example water and oil (O), consists essentially of adjacent monolayers of the two liquids, and the interfacial tension is the sum of the two individual tensions. Both liquids have, when separated, their normal surface tension, but when in contact the molecules in the water monolayer are subject to forces from both the bulk water and the oil molecules, hence the tension is diminished – similarly for the oil monolayer. The magnitude of the oil-water force is a function of the dispersion forces of the oil and water molecules; it has been shown [11] that the geometric mean of the contributions to surface tension of these forces $({}^d\gamma^O \; {}^d\gamma^W)^{1/2}$, accurately describes the force of interaction. The tension in the oil layer is reduced by the same quantity, hence

$$\gamma^{OW} = \gamma^{OV} + \gamma^{WV} - 2\sqrt{({}^d\gamma^O \; {}^d\gamma^W)} \tag{1.5}$$

where γ^{OV} and γ^{WV} are the surface tensions of oil and water respectively.

In general for any two immiscible liquids 1 and 2, an estimate for ${}^d\gamma^1$ (or ${}^d\gamma^2$) may be obtained by summing the pair potentials of all the surface volume elements with all the volume elements below the surface, giving

$$^\mathrm{d}\gamma^1 = - \frac{\pi N_1{}^2 \alpha_1{}^2 I_1}{8r_{11}{}^2} \tag{1.6}$$

where α_1 is the polarisability, I_1 the ionization potential, N_1 the number of molecules of type 1 per unit volume, and r_{11} the intermolecular distance. For water a value of 25.4 mJ m^{-2} is obtained for the dispersion force contribution to the surface tension. Similarly, for the interaction between dissimilar phases, at an interface, with volume elements of such a size that those in the surface have 12 nearest neighbours, then the energy of interaction of the volume elements in the surface of the phase of molecules 1 is given by

$$^\mathrm{d}\gamma^{12} = - \frac{\pi N_1 N_2 \alpha_1 \alpha_2}{4r_{12}{}^2} \quad \frac{I_1 I_2}{I_1 + I_2} \quad . \tag{1.7}$$

Using the geometric mean relationship,

$$^\mathrm{d}\gamma^{12} = (^\mathrm{d}\gamma^1 \, ^\mathrm{d}\gamma^2)^{\frac{1}{2}} = - \frac{\pi N_1 N_2 \alpha_1 \alpha_2}{8r_{11} r_{22}} \quad \frac{I_1 I_2}{\sqrt{(I_1 I_2)}} \quad . \tag{1.8}$$

Comparing these last two equations shows that except for large differences between r_{11} and r_{22} and between I_1 and I_2, the geometric mean is a satisfactory average for the calculation of intermolecular forces at interfaces. To illustrate the point, Fowkes [10] calculated an average value of 21.8 ± 0.7 mN m^{-1} for $^\mathrm{d}\gamma^\mathrm{W}$, using experimental values of $\gamma^{1\mathrm{V}}$, $\gamma^{2\mathrm{V}}$ and γ^{12} for eight different saturated hydrocarbons with water, in equation (1.5). Similarly for mercury with ten hydrocarbons, $^\mathrm{d}\gamma^\mathrm{M} = 200 \pm 7$ mN m^{-1}. For the mercury/water interface equation (1.5) gives $\gamma^{12} = 424.8 \pm 4.4$ mN m^{-1}, as compared with the experimental values of 426–7 mN m^{-1}; evidently the interaction between these two liquids is almost entirely the result of dispersion forces.

So it would appear that dividing up the surface tension into components, as originally proposed by Fowkes, is a valuable concept. In fact he distinguished at least seven components in the interaction across an interface, including dispersion forces, hydrogen bonds, dipole-dipole interactions, dipole-induced dipole interactions, Π-bonds, donor-acceptor bonds, and electrostatic interactions; the first is usually the dominant term But it should be pointed out that terms may not be independent, and Good [12] notes that the interaction term between $^\mathrm{d}\gamma$ (dispersion) and $^\mathrm{p}\gamma$ (dipole-dipole), cannot be neglected when at least one dipole moment in the system is greater than about 1.5 Debye units.

1.4 ATTRACTIVE FORCES AT INTERFACES
Let us now consider the attractive forces that exist at interfaces, taking first the interaction between a molecule or atom and a solid surface, and then that between two bulk phases. No new forces are involved — just the summation of elementary interactions.

The total attractive energy between an atom and a solid of infinite extent and depth, may be obtained by assuming simple additivity of forces, and summing over all atom-atom interactions, expressed as $U = \Sigma \, U_i$, where U_i is the potential energy of interaction between the external atom and the ith atom of the solid, assuming $U = 0$ at infinite separation. Normally U_i is assumed to be of the form $U = -C/r^6$, where C is the London constant. London replaced the sum by an integral

$$U = -\int Cr^{-6} N \mathrm{d}v \tag{1.9}$$

where N is the number of atoms in unit volume of the solid and $\mathrm{d}v$ is a volume element of solid at a distance r from the atom. If N is assumed constant

$$U = -N\pi C/6d^3 \tag{1.10}$$

where d is the shortest distance between atom and surface. These assumptions are only reasonable if d is much larger than the space between atoms in the solid, so are not applicable when the atom is adsorbed on the surface. However, the principle has been applied successfully to the simplest case of the non-polar spherical molecule or atom interacting with a non-ionic surface. For other more complex systems such as metals, conductors etc. further terms are included; these were summarised in 1962 by Young and Crowell [13]. More recently wave-mechanical methods have been developed.

To obtain the attractive potential energy U_A between two slabs of solid separated by a distance d in vacuum, a further integration is required over the depth of the second slab, which leads to an energy proportional to d^2

$$U_A = -A/12\pi d^2 \tag{1.11}$$

where A is the **Hamaker constant** (after its user), equal to $\pi^2 N^2 C$. For two solids α and β we have

$$U_A^{\alpha\beta} = -A^{\alpha\beta}/12\pi d^2 \tag{1.12}$$

and here $A^{\alpha\beta} = \pi^2 N_\alpha \, N_\beta \, C_{\alpha\beta}$, where N_α and N_β are the number densities of atoms in α and β respectively, and $C_{\alpha\beta}$ is the London constant for the interaction of an α atom with a β atom. This approach applies equally to the interaction between two liquids, but the pairwise summation cannot be expected to hold accurately at values of d of the order of atomic or molecular dimensions.

An alternative to this 'microscopic' approach was introduced by Lifshitz [14] in 1955, and it is, in contrast, called the 'macroscopic' approach. The fluctuating electric field that arises from the distribution of charge in individual atoms is seen to extend beyond the surface of the bulk phase, and gives rise to fluctuations in a nearby surface. Field strengths are calculated from dielectric susceptibilities. Attractive forces, both retarded and non-retarded, are obtained from the theory, provided that the basic physical data are available; the results are comparable with those from the pairwise summation theory. There is a difference – Lifshitz's theory includes all interaction frequencies, not simply that in the ultraviolet which is the case for the London energy. Again, it is not really valid for small values of d.

Finally we will consider the effect of interposing a medium on the long-range interaction between the two slabs of solid. The Hamaker constant (equation (1.11)) must be modified to take account of the fact that the dispersion forces are transmitted through the intervening medium. For a single substance the value of A can be calculated by summing the pair potentials (μ) between volume elements.

$$\mu_{11} = -\frac{3\alpha^2 I_1}{4r_{11}^6} \tag{1.13}$$

and the summation gives

$$A_1 = \tfrac{3}{4}\,\pi^2 N_1^2 \alpha_1^2 I_1 \quad . \tag{1.14}$$

When solid 2 is immersed in substance 1, Hamaker showed that

$$A_{12} = \sqrt{A_1}^2 - 2\sqrt{(A_1 A_2)}\,\frac{2\sqrt{(I_1 I_2)}}{I_1 + I_2} + \sqrt{A_2}^2 \quad , \tag{1.15}$$

and when I_1 and I_2 are not too different

$$A_{12} = (\sqrt{A_1} - \sqrt{A_2})^2 \quad . \tag{1.16}$$

We have already seen (equation (1.6)) that the summation of pair potentials leads to a value for ${}^d\gamma^1$, so substitution of equation (1.6) into equation (1.14) gives

$$A_1 = 6\pi r_{11}^2\,{}^d\gamma^1 \tag{1.17}$$

and

$$A_{12} = 6\pi r_{11}^2\,(\sqrt{{}^d\gamma^1} - \sqrt{{}^d\gamma^2})^2 \quad . \tag{1.18}$$

Therefore Hamaker constants can be estimated from experimental values of the dispersion force contribution to the surface tension. Fowkes [11] published calculated values of A_{12}, using this approach, for water with solids that have similar sized volume elements to that of water, for example oxide ions, CH_2 or CH groups, and metal atoms. Table 1.1 shows how these values of A_{12} vary with $^d\gamma^2$, for a variety of solids immersed in water. But the values cannot be accurate, as Gregory [15] points out, since the interactions considered are only those for molecules very close to the interface; the summation should be carried out over all pairs of interacting molecules, but the distances involved are not accurately known since the molecular structure of liquids is poorly understood. There are also other indirect methods for deriving values of A_{12}, for example from studies of the flocculation of colloidal particles, but so many other parameters are involved, some quite ill-defined, that any agreement between results should not be seen to be proof that the theory is adequate. In a large number of cases the values obtained from colloid chemistry and surface tension measurements deviate substantially from those calculated from dispersion force theory.

Values of Hamaker constants may be obtained from London's treatment of dispersion forces, as well as by using the macroscopic approach of Lifshitz. Gregory [15] and Visser [16] compared the two methods, and found reasonable agreement provided that optical dispersion data are used in the calculations. Furthermore, the agreement is extremely good for materials separated by water.

Table 1.1

A_{12} for various substances immersed in water at 20°C (taken from reference [11])

	$^d\gamma^2$/mJ m^{-2}	A_{12}/J
Polyhexafluoropropylene	18	2×10^{-22}
Paraffin wax	25.5	2×10^{-22}
Polyethylene	35	2×10^{-21}
Polystyrene	44	5×10^{-21}
Copper	60	1.4×10^{-20}
Silver	76	2.5×10^{-20}
Anatase (TiO_2)	91	3.5×10^{-20}
Iron	98	4×10^{-20}
Iron (III) oxide	107	4.5×10^{-20}
Graphite	110	5×10^{-20}
Tin (IV) oxide	118	5.5×10^{-20}
Silica	123	6×10^{-20}
Rutile (TiO_2)	143	8×10^{-20}
Mercury	200	1.3×10^{-19}

1.5 MEASUREMENT OF ATTRACTIVE FORCES AT INTERFACES

Several indirect methods of determining attractive energy terms have already been mentioned, and they all involve parameters which are uncertain. For example, interpretation of experiments on the kinetics of flocculation of colloidal particles requires an understanding of the repulsive forces associated with overlap of the electrical double layers. Furthermore the use of surface tension data assumes information on packing of molecules. So attempts have been made to measure directly the forces between solid bodies. These started in Moscow in 1951 (Derjaguin and Abrikosova [17]), and have since developed into very sophisticated experimentation. Retarded forces at large distances (~100 nm) were first observed by Derjaguin, and the transition from non-retarded to retarded behaviour was demonstrated by Israelachvili and Tabor [18] in 1972 using crossed cylinders of mica. Mica is a good substance to use in such experiments since it can be made molecularly smooth. Excellent agreement with Lifshitz's theory was also found. More recently the forces between two mica surfaces in aqueous electrolyte solutions have been measured in the range 0 – 100 nm by Israelachvili and Adams [19]. The attractive force was separated out by measurements at high electrolyte concentration, at which the repulsion due to electric double layer effects is weak. Up to about 6.5 nm the force is non-retarded, with $A_{12} = (2.2 \pm 0.3) \times 10^{-20}$J (cf Table 1.1), and at larger distances retardation effects are manifest. It is interesting to note that the van der Waals forces between two mica surfaces in aqueous solutions are much reduced in magnitude from those in air, and that retardation effects set in earlier.

The experimentation was reviewed by Israelachvili and Tabor [20] in 1973, and the uncertainties in the interpretation of data obtained for small separation distance are discussed in Mahanty and Ninham's book (1976) on dispersion forces [21].

1.6 REFERENCES

[1] T. Young, (1805) *Trans. Roy. Soc.* (London), **95**, 65.
[2] J. W. Gibbs, (1906) *Collected Works,* Vol. 1, Longmans, London (Dover, 1961).
[3] A. W. Adamson, (1976) *Physical Chemistry of Surfaces,* 3rd ed. Wiley p. 55.
[4] E. R. S. Winter, (1950) *Disc. Faraday Soc.,* **8**, 231.
[5] E. A. Moelwyn-Hughes, (1961) *Physical Chemistry,* 2nd ed. Pergamon Press, Oxford Ch. 7.
[6] M. Davies, (1965) *Some Electrical and Optical Aspects of Molecular Behaviour,* Pergamon Press, Oxford Ch. 7.
[7] D. Tabor, (1969) *Gases, Liquids and Solids,* Penguin Books, Harmondsworth Ch. 2.
[8] G. Mie, (1903) *Ann. der Physik,* **11**, 657.

[9] S. Ross, Private communication.

[10] F. M. Fowkes, (1963) *J. Phys. Chem.,* **67**, 2538.

[11] F. M. Fowkes, (1964) *Ind. Eng. Chem.,* **56**, No 12, 40.

[12] R. J. Good,(1977) *J. Colloid Interface Sci.,* **59**, 398.

[13] D. M. Young and A. D. Crowell, (1962) *Physical Adsorption of Gases,* Butterworths, London Ch. 2.

[14] E. M. Lifshitz, (1955) *Zh. Eksper. i Teor. Fiz.,* **29**, 94.

[15] J. Gregory, (1970) *Adv. Colloid Interface Sci.,* **2**, 418.

[16] J. Visser, (1972) *Adv. Colloid Interface Sci.,* **3**, 331.

[17] B. V. Derjaguin and I. I. Abrikosova, (1956) *J. Exp. Theor. Phys.,* **30**, 993.

[18] J. N. Israelachvili and D. Tabor, (1972) *Proc. Roy. Soc. A,* **331**, 19.

[19] J. N. Israelachvili and G. E. Adams, (1978) *J. Chem Soc. Faraday I,* **74**, 975.

[20] J. N. Israelachvili and D. Tabor, (1973) *Prog. Surface Membrane Sci.,* 7, 1.

[21] J. Mahanty and B. W. Ninham, (1976) *Dispersion Forces,* Academic Press, New York Ch. 3.

CHAPTER 2

Thermodynamic description of an interface

2.1 INTRODUCTION

It is not an adequate description of the junction between two homogeneous bulk phases to describe it as a simple two-dimensional plane without thickness. Because of the finite range of action of intermolecular forces the interface can more properly be regarded as a region of finite thickness across which the energy, density or any other thermodynamic property changes gradually. Since this region has both area and thickness it may be considered as an interphase, and this may exist in the solid, liquid or gaseous states; they are usually referred to as **two-dimensional** phases, since the thickness parameter cannot be varied at will by the experimenter.

2.2 DEFINITION OF AN INTERFACE

The original treatment of J. Willard Gibbs which appeared in *Influence of Surfaces of Discontinuity upon the Equilibrium of Heterogeneous Masses*, is rather abstract. However, the approach of van der Waals and Bakker, discussed in Guggenheim's book (see bibliography) is equivalent and more easily understood.

Consider two homogeneous bulk phases, α and β, and separating them an interfacial layer or interphase s (Fig. 2.1). The boundary between the interphase and the bulk phase α is the plane AA', and that between the interphase and the bulk phase β is the plane BB'.

The properties of the interphase are assumed to be uniform in any plane parallel to AA' or BB', but not in any other plane in the interphase. At and near the plane AA' the properties of s are indentical with those of the bulk phase β. However, moving from AA' to BB' within the region s represents a gradual change in the properties of the interphase, from those corresponding to phase α to those corresponding to phase β.

It is usual, although not essential, to imagine the interphase as being submicroscopic in thickness, say 1 to 10 nm.

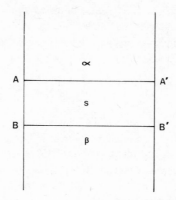

Fig. 2.1 – Definition of an interphase

2.3 INTERFACIAL TENSION AT A PLANE SURFACE

The interface as just defined can be treated as a thermodynamic system, either **open** or in certain cases **closed**. A thermodynamic system is defined as open if both energy and matter may be transferred across the boundaries of that system. The system is said to be closed if no matter is allowed to move across the boundary, or if neither matter nor energy is transferred it is said to be **isolated**. Normally, in this chapter, the interphase will be considered as an open system unless otherwise stated, and the only quantity which requires special consideration is the interfacial tension.

In a bulk phase the force across any unit area is equal in all directions, as is the pressure. But in the interphase the force is not the same in all directions. However, if a plane of unit area is chosen parallel to AA' or BB', the force across the plane is the same for any position of the plane whether it lie in α, β or s, since hydrostatic changes are assumed negligible; this is the pressure, P.

If a plane is chosen perpendicular to AA', then the situation is somewhat different. Let this plane be represented by a rectangle of height h, parallel to AB, and of length l perpendicular to the plane of the diagram (see Fig. 2.1). The force across such a plane will be equal to

$$Phl - \gamma l$$

where γ is the interfacial tension, this representing the combined effects of the pressure and interfacial tension forces. If h is chosen to extend exactly from AA' to BB', that is it represents the interphase thickness τ, then the force across the rectangle will be

$$P\tau l - \gamma l,$$

the difference in sign being the difference between work done on the system by the external pressure, and work done by the system involving the interfacial tension forces. The relationship between this thickness and the interfacial volume V^s and area A, is

$$V^s = \tau A \quad .$$

Suppose that the area is increased by dA, the volume by dV^s, and the thickness by $d\tau$, the material content remaining unaltered. The work done on the interphase by the forces across the plane AA' and BB' is $-PAd\tau$. The work done by forces parallel to the planes AA' and BB' is independent of the shape of the perimeter, which can be assumed for simplicity to be a rectangle. Consequently the work done on s by the latter forces is then

$$-(P\tau - \gamma)dA$$

and the total work done on s will be the sum

$$
\begin{aligned}
&-PA\ d\tau - (P\tau - \gamma)dA \\
&= -P(Ad\tau + \tau dA) + \gamma dA \\
&= -PdV^s + \gamma dA \quad .
\end{aligned}
\tag{2.1}
$$

This last expression is the analogous work term for an interphase which corresponds to the three-dimensional $-PdV$ for a bulk phase.

This changed work term is often difficult to comprehend but can perhaps be illustrated in a non-rigorous way as follows. The interfacial tension has the basic unit of force per unit length, that is, $N\ m^{-1}$. This could be rewritten as $N\ m\ m^{-2}$, but the quantity $N\ m$ is by definition the Joule, J. Thus the interfacial tension can also be considered as an energy per unit area, $J\ m^{-2}$. Clearly therefore the term $\gamma\ dA$ has the unit J, the same as the usual work term PdV.

The first law of thermodynamics states that

$$dU = q + w \tag{2.2}$$

so that if for the moment we put $q = 0$, the energy change of system would then be

$$dU = -PdV^s + \gamma dA \quad . \tag{2.3}$$

If we apply this to an interphase treated as a closed system and imagine an increase in both V^s and A, then both dV^s and dA must be positive. Thus an increase in volume will result in work being done against the external pressure, and a decrease in internal energy of $-PdV^s$; on the other hand — since molecules in the surface have greater energy than those in the bulk, and we have already stated that dA is positive $-$ γ dA must therefore represent an energy increase of the system, and the respective signs of the terms in equation (2.1) are clearly understandable.

From now on, for simplicity, superscript s denoting the interphase is omitted.

2.4 THE FREE ENERGY OF AN INTERFACE
2.4.1. The 'classical' approach, as adopted by Harkins and others

Let us consider a closed plane interface, that is of fixed composition, with γ, the tangential stress, as an intensive variable, that is to say, one in which the value for the system as a whole is not the sum of its values at various points within that system (other examples of intensive variables having the same value at any point in the system are temperature and pressure).

Making the substitution $q_{\text{reversible}} = TdS$ in equation (2.2) and using equation (2.1) we have

$$dU = T\text{d}S - P\text{d}V + \gamma\text{d}A \quad . \tag{2.4}$$

The usual thermodynamic function definitions are

enthalpy $\qquad H = U + PV$ $\qquad\qquad\qquad\qquad\qquad$ (2.5)

available energy $\quad A = U - TS$ $\qquad\qquad\qquad\qquad\qquad$ (2.6)

Gibbs free energy $\quad G = H - TS = U + PV - TS \quad .$ $\qquad\qquad$ (2.7)

Let us first consider the available energy A as defined above. Differentiation yields

$$dA = dU - T\text{d}S - S\text{d}T \tag{2.8}$$

which on substitution yields

$$dA = -S\text{d}T - P\text{d}V + \gamma\text{d}A \tag{2.9}$$

and consequently

$$\left[\frac{\partial A}{\partial A}\right]_{T,V} = \gamma \tag{2.10}$$

$$\left[\frac{\partial A}{\partial T}\right]_{V,A} = -S \tag{2.11}$$

$$\left[\frac{\partial A}{\partial V}\right]_{T,A} = -P \quad . \tag{2.12}$$

At constant temperature

$$dA = dU - TdS$$

$$= -w_{max}$$

$$= -PdV + \gamma dA \tag{2.13}$$

if no other work than that associated with volume and area changes is involved.

The Gibbs function for a system containing an interface is not as well behaved. For bulk phase systems it is normal to consider that the change in the Gibbs function represents the net work of the system. When a system which includes a surface is considered, the 'classical' approach no longer leads to this simple equality. Differentiation of equation (2.7) yields

$$dG = dU + PdV + VdP - TdS - SdT \tag{2.14}$$

which on substitution gives

$$dG = -SdT + VdP + \gamma dA \tag{2.15}$$

hence

$$\left[\frac{\partial G}{\partial A}\right]_{T,P} = \gamma \tag{2.16}$$

$$\left[\frac{\partial G}{\partial T}\right]_{P,A} = -S \tag{2.17}$$

$$\left[\frac{\partial G}{\partial P}\right]_{T,A} = V \tag{2.18}$$

and at constant temperature and pressure

$$dG = dU + PdV - TdS$$

$$= -w_{max} + PdV$$

$$\neq \text{net work} \tag{2.19}$$

since the contribution to the work term of $\gamma \mathrm{d}A$ is not in the equation. The fact that the free energy change $\mathrm{d}G$ is no longer equal to the net work, is rather unfortunate. Obviously it would be better if the free energy change $\mathrm{d}G$ were always equal to the net work regardless as to whether the system contained an interface or not. It is this abnormality in behaviour of the Gibbs function which has led to the redefinitions of both the enthalpy and Gibbs function as described in the following section.

2.4.2 The methods of Guggenheim and Hill (see Bibliography)

The essential difference between the approach in this section and that in the previous section lies in the definition of enthalpy. Using the new definition given below this will now be written as H.

Substitution for the work term in the first law expression (2.2) yields

$$\mathrm{d}U = q - P\mathrm{d}V + \gamma\mathrm{d}A \tag{2.20}$$

which may be rearranged to give

$$q_{P,\gamma} = \mathrm{d}U + P\mathrm{d}V - \gamma\mathrm{d}A$$

$$= \mathrm{d}H \tag{2.21}$$

Hence we may define

$$H = U + PV - \gamma A \quad . \tag{2.22}$$

Remembering the expression (2.3) we may therefore define

$$A = U - TS \tag{2.6}$$

$$G = H - TS \quad . \tag{2.23}$$

Thus the definition of A is unaltered from what we have termed the classical approach, whilst G is a different quantity from G.

Since the definition of the available energy has not been altered, the equation previously derived to illustrate the relationship to the maximum work still holds. The same is not true in the case of the redefined variable G. If equation (2.23) and (2.22) are combined

$$G = U + PV - \gamma A - TS \tag{2.24}$$

then by differentiation

$$\mathrm{d}G = \mathrm{d}U + P\mathrm{d}V + V\mathrm{d}P - \gamma\mathrm{d}A - A\mathrm{d}\gamma - T\mathrm{d}S - S\mathrm{d}T$$

which under conditions of constant T, P and γ yields

$$\mathrm{d}G_{T,P,\gamma} = \mathrm{d}U + P\mathrm{d}V - \gamma\mathrm{d}A - T\mathrm{d}S$$

$$= -(w_{\mathrm{max.}} - P\mathrm{d}V + \gamma\mathrm{d}A) \tag{2.25}$$

which represents the net work as would be usual for the Gibbs function. By similar methods to those previously employed it can be shown that

$$\left[\frac{\partial G}{\partial T}\right]_{P,\gamma} = -S \tag{2.26}$$

$$\left[\frac{\partial G}{\partial P}\right]_{T,\gamma} = V \tag{2.27}$$

$$\left[\frac{\partial G}{\partial \gamma}\right]_{T,P} = -A \tag{2.28}$$

and

$$G = H + T \left[\frac{\partial G}{\partial T}\right]_{P,\gamma} \tag{2.29}$$

2.4.3 Treatment of an open system by 'classical' methods

For an open system with varying composition, where μ_i and n_i are the chemical potential and number of moles of type i, then

$$\mathrm{d}U = T\mathrm{d}S - P\mathrm{d}V + \gamma\mathrm{d}A + \Sigma_i \mu_i \mathrm{d}n_i \tag{2.30}$$

and consequently it may be deduced

$$\mathrm{d}H = \mathrm{d}U + P\mathrm{d}V + V\mathrm{d}P$$

$$= T\mathrm{d}S + V\mathrm{d}P + \gamma\mathrm{d}A + \Sigma_i \mu_i \mathrm{d}n_i \tag{2.31}$$

$$\mathrm{d}A = \mathrm{d}U - T\mathrm{d}S - S\mathrm{d}T$$

$$= -S\mathrm{d}T - P\mathrm{d}V + \gamma\mathrm{d}A + \Sigma_i \mu_i \mathrm{d}n_i \tag{2.32}$$

$$dG = dH - TdS - SdT$$

$$= -SdT + VdP + \gamma dA + \Sigma_i \mu_i dn_i \tag{2.33}$$

Therefore

$$\mu_i = \left[\frac{\partial U}{\partial n_i}\right]_{n_j, S, V, A}$$

$$= \left[\frac{\partial H}{\partial n_i}\right]_{n_j, S, P, A}$$

$$= \left[\frac{\partial A}{\partial n_i}\right]_{n_j, T, V, A}$$

$$= \left[\frac{\partial G}{\partial n_i}\right]_{n_j, T, P, A} . \tag{2.34}$$

Integration of the above equations at constant intensive variables

$$U = TS - PV + \gamma A + \Sigma_i \mu_i n_i \tag{2.35}$$

$$H = TS + \gamma A + \Sigma_i \mu_i n_i \tag{2.36}$$

$$A = -PV + \gamma A + \Sigma_i \mu_i n_i \tag{2.37}$$

$$G = \gamma A + \Sigma_i \mu_i n_i \tag{2.38}$$

the latter equation leading to an unexpected result for G, in which a term γA is included in addition to the expected sum, and would lead to a peculiar partial expression for the chemical potential.

2.4.4 The methods of Guggenheim and Hill applied to an open system

This approach gives rise to more readily understandable results. The starting point is again the equation for the total energy of an open system with varying composition which has previously been stated as equation (2.30):

$$dU = TdS - PdV + \gamma dA + \Sigma_i \mu_i dn_i$$

and from which it may be deduced that

$$dH = dU + PdV + VdP - \gamma dA - Ad\gamma$$

$$= TdS + VdP - Ad\gamma + \Sigma_i \mu_i dn_i \qquad (2.39)$$

$$dA = dU - TdS - SdT$$

$$= -SdT - PdV + \gamma dA + \Sigma_i \mu_i dn_i$$

which is identical with the equation deduced by classical methods, equation (2.32). However, the analogous expression to equation (2.33) is different:

$$dG = dH - TdS - SdT$$

$$= -SdT + VdP - Ad\gamma + \Sigma_i \mu_i dn_i \qquad (2.40)$$

which equations give rise to the expressions for the chemical potential

$$\mu_i = \left[\frac{\partial U}{\partial n_i} \right]_{n_j, S, V, A}$$

$$= \left[\frac{\partial H}{\partial n_i} \right]_{n_j, S, P, \gamma}$$

$$= \left[\frac{\partial A}{\partial n_i} \right]_{n_j, T, V, A}$$

$$= \left[\frac{\partial G}{\partial n_i} \right]_{n_j, T, P, \gamma} \qquad (2.41)$$

all of which would appear consistent with the results for a bulk phase alone. By integration of the above equation expressions for the functions themselves may be obtained. However, since the expressions for dU and dA are identical in both cases the integrated expressions must be identical and thus yield the equations (2.35) and (2.37), but the results for the remaining functions are

$$H = TS + \Sigma_i \mu_i n_i \qquad (2.42)$$

and

$$G = \Sigma_i \mu_i n_i \quad . \qquad (2.43)$$

This latter result is important since it is of the form familiar from the treatment of bulk phases, and hence is of universal applicability, whereas the function G gives rise to the peculiar equation (2.38) for a surface phase.

The methods of Guggenheim and Hill, although involving re-definition of H and G, lead to a more consistent treatment of interphases than what may be termed classical methods. Using their approach the significance of thermodynamic parameters is no longer altered by the presence or absence of an interphase in the system. In fact the definitions of H and G are precise generalised definitions, and H and G should be regarded as specific cases for bulk phases in the absence of an interphase.

2.5 THE GIBBS-DUHEM EQUATION FOR A INTERPHASE
We have previously deduced that

$$dG = -SdT + VdP - Ad\gamma + \Sigma_i \mu_i dn_i \qquad (2.40)$$

and if equation (2.43) is differentiated

$$dG = \Sigma_i \mu_i dn_i + \Sigma_i n_i d\mu_i \quad . \qquad (2.44)$$

Thus by subtracting these two equations

$$SdT - VdP + Ad\gamma + \Sigma_i n_i d\mu_i = 0 \qquad (2.45)$$

which is the analogue for an interphase of the usual form of the Gibbs-Duhem equation for a bulk phase which merely omits the term $Ad\gamma$.

2.6 THE GIBBS ADSORPTION ISOTHERM
The quantity of the components adsorbed at the interface is obviously a significant parameter, and particularly of interest is the relationship between the extent of adsorption and the interfacial tension. Since it is difficult to make systematic measurements when many factors vary simultaneously, it is convenient to investigate the relationship between the extent of adsorption and interfacial tension whilst the temperature is held constant; hence the **adsorption isotherm**.

The original derivation of what is variously known as either the Gibbs adsorption isotherm or the Gibbs equation is somewhat more difficult to follow than that starting from the Gibbs-Duhem equation.

The Gibbs-Duhem relation for an interphase under conditions of constant temperature and pressure gives

$$- Ad\gamma = \Sigma_i n_i d\mu_i \quad . \qquad (2.46)$$

If we now write

$$\Gamma_i = \frac{n_i}{A} \tag{2.47}$$

where Γ_i represents the amount of species i in unit area of the interphase, then we have the Gibbs adsorption isotherm

$$-d\gamma = \Sigma_i \, \Gamma_i d\mu_i \tag{2.48}$$

where the quantity Γ_i requires exact definition which is dependent on the choice of boundary or dividing surface for the interphase.

2.7 CHOICE OF A DIVIDING SURFACE

It is easiest to illustrate the use of such an equation by references to a two-component liquid mixture, and the example frequently chosen to illustrate the point is that of ethanol and water, where water is the solvent and ethanol the solute.

The Gibbs adsorption isotherm for a two-component system may be written as

$$-d\gamma = \Gamma_1 \, d\mu_1 + \Gamma_2 \, d\mu_2 \tag{2.49}$$

where the subscript 1 refers to the solvent and subscript 2 to the solute. Whilst recognising that the interfacial region is best regarded as an interphase, as shown in earlier sections of this chapter, there is the alternative mathematical model of considering the interphase as a plane of infinitesimal thickness situated between AA' and BB' of Fig. 2.1. This dividing surface may be considered to be positioned so as to give rise to simplification of equation (2.49). Gibbs defined the position of the dividing surface such that the surface excess of solvent is zero, and hence we may write

$$-d\gamma = {}^1\Gamma_2 \, d\mu_2 \tag{2.50}$$

where ${}^1\Gamma_2$ is now termed the **surface excess** of the solute with the dividing surface so defined. The significance of this may be seen from the Fig. 2.2. To quote from Gibbs' original paper:

It is generally possible to place the dividing surface so that the total quantity of any desired component in the vicinity of the surface of discontinuity shall be the same as if the density of that component were uniform on each side quite up to the dividing surface. In other words we may place the dividing surface so as to make any one of the quantities Γ_1, Γ_2, etc., vanish.

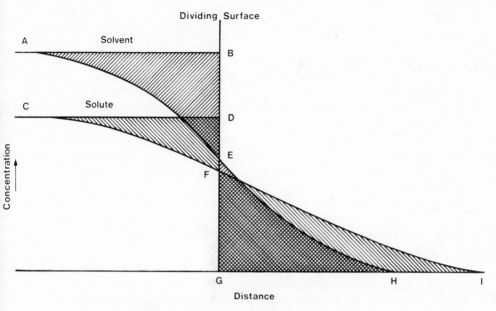

Fig. 2.2 – Definition of a dividing surface

Thus the dividing surface in Fig. 2.2 is such that the surface excess of solvent is zero, that is the shaded area ABE will be equal to the shaded area EGH. The difference between areas FGI and CDF will give *the surface excess of solute* $^1\Gamma_2$, *which is defined as the algebraic excess of solute in unit cross-section of the surface region over that which would be present in a bulk region containing the same number of moles of solvent as does the section of the surface region.*

A corresponding symmetrical definition of $^2\Gamma_1$ can be written where Γ_2 is equal to zero, and $^2\Gamma_1$ represents the excess of solvent in unit cross-section of the surface region over that which would be present in a bulk region containing the same number of moles of solute as does the section of the surface region.

The above is not the only way of locating the dividing surface, and several alternative definitions are possible. The dividing surface could be placed such that the algebraic sum areas in Fig. 2.2 to the right of the dividing surface is equal to the sum of the areas to the left of the dividing surface. The surface excess defined in this manner may be written $^n\Gamma_i$, which represents the excess of the ith component in unit cross-section of the surface region over the moles that would be present in a bulk region containing the same total number of moles as does the section of the surface region.

A quantity $^m\Gamma_i$ may be defined as the excess of the ith component in the region of the surface over the number of the moles that would be present in a bulk region of the same total mass as the surface region.

Instead of using the total mass in the definition the total volume may be used. Thus $^v\Gamma_i$ is the excess of the ith component in the region of the surface over the number of moles that would be present in a bulk region of the same total volume as the surface region.

The exact significance of these various definitions of surface excess is somewhat difficult to visualise, and a numerical example is of some assistance. Consider a bulk of solution of ethanol in water, where the mole fraction of each component is 0.5. In order to make the arithmetic easy to follow we will use numbers which are of the right order of magnitude but not experimentally exact, but they will serve the purpose of illustration well enough. If a sample of approximately 0.1 cm^3 of the surface region is taken containing 1 mmol of water and 1.016 mmol of ethanol from a surface area of 2 m^2, it is probable that some of the bulk solution is included in this sample together with the whole of the surface region. In order to calculate the various surface excesses the following data will also be required, namely that the relative molecular weights of water and ethanol are 18 and 46, and that the value of the respective molar volumes are 18 and 58 cm^3 mol^{-1}.

The quantity $^1\Gamma_2$ is the excess which 1.016 mmol of ethanol represents over the quantity present in the bulk associated with the same amount of water as in the surface. The surface sample contains 1 mmol of water, and in the bulk 1 mmol of water is associated with 1 mmol of ethanol. Thus the surface excess is equal to 1.016-1 = 0.016 mmol for an area of 2 m^2. Thus $^1\Gamma_2 = 0.016/2$ mmol m^{-2} = 8 μmol m^{-2}.

The symmetrical definition of $^2\Gamma_1$ can be calculated as well. If the surface phase sample contains 1.016 mmol of ethanol the surface excess of water will be the excess of 1 mmol of water over the amount of water associated with 1.016 mmol of ethanol in the bulk solution. The surface excess is therefore equal to 1-1.016 = −0.016 mmol for an area of 2 m^2. Thus $^2\Gamma_1 = -0.016/2$ mmol m^{-2} = −8 μmol m^{-2}.

The remaining three definitions are calculated somewhat differently. Firstly we have the quantities $^n\Gamma_1$ and $^n\Gamma_2$, in which the bulk region for comparison contains the same total number of moles as the surface region. If the surface region contains 1 mmol of water and 1.016 mmol of ethanol, then the bulk region for comparison will contain 1.008 mmol water and 1.008 mmol of ethanol. The surface excess of ethanol is 1.016-1.008 = 0.008 mmol and that of water is 1-1.008 = −0.008 mmol for an area of 2 m^2. Thus $^n\Gamma_1 = -0.008/2$ mmol m^{-2} = −4 μmol m^{-2} and $^n\Gamma_2 = 0.008/2$ mmol m^{-2} = 4 μmol m^{-2}.

The next definition relates to the quantities $^m\Gamma_1$ and $^m\Gamma_2$ in which case the total mass is the basis of comparison. The sample of surface has a total mass of $(10^{-3} \times 18) + (1.016 \times 10^{-3} \times 46) = 0.0647$ g. This amount of bulk solution contains 18.2 mg of water and 46.5 mg of ethanol, that is 1.011 mmol of water and 1.011 mmol of ethanol. The surface excess of ethanol will be 1.016-1.011 mmol = 0.005 mmol and that of water 1-1.011 = −0.011 mmol for an area of 2 m^2. Thus $^m\Gamma_1 = -5.5$ μmol m^{-2} and $^m\Gamma_2 = 2.5$ μmol m^{-2}.

The remaining definition is on the basis of comparison of equal volumes of surface and bulk regions. The total volume of the surface region is $(10^{-3} \times 18) + (1.016 \times 10^{-3} \times 58) = 0.0769$ cm^3 and this volume of bulk solution will contain 1.012 mmol of ethanol and of water. The surface excess of ethanol will be $1.016 - 1.012 = 0.004$ mmol and that of water $1 - 1.012 = -0.012$ mmol for an area of 2 m^2. Thus $^v\Gamma_1 = -6$ μmol m^{-2} and $^v\Gamma_2 = 2$ μmol m^{-2}.

We can regard the term surface excess as a means of expressing surface concentrations in a way that relates such a parameter to the bulk phase composition with which the interphase is in equilibrium. The thermodynamic approach which we have outlined systematises the possible definitions, and the relationship of such quantities to other basic thermodynamic functions.

2.8 THE RELATIONSHIP BETWEEN THE VARIOUSLY DEFINED SURFACE EXCESSES

The Gibbs-Duhem relationship for a bulk phase may be written

$$S dT - V dP + \Sigma_i n_i d\mu_i = 0$$

which at constant temperature and pressure reduces to

$$\Sigma_i n_i d\mu_i = 0$$

which may be written for a two-component system as

$$n_1 d\mu_1 + n_2 d\mu_2 = 0$$

or if mole fractions are employed

$$x_1 d\mu_1 + x_2 d\mu_2 = 0 \quad .$$

Substitution in the Gibbs adsorption isotherm as written in equation (2.49) yields

$$-d\gamma = -\Gamma_1 \frac{x_2}{x_1} d\mu_2 + \Gamma_2 d\mu_2 \quad .,$$

Therefore

$$-\frac{d\gamma}{d\mu_2} = \Gamma_2 - \frac{x_2}{x_1} \Gamma_1 \quad . \tag{2.51}$$

Since the quantity $-d\gamma/d\mu_2$ is independent of the choice of dividing surface, then it follows that the expression on the right-hand side of equation (2.51) must also be independent.

It can be seen from the numerical example that the quantities $^n\Gamma_i$, $^m\Gamma_i$ and $^v\Gamma_i$ may be related by equations of the type

$$f_1 Q_1 + f_2 Q_2 = 0 \tag{2.52}$$

where the factors f_i are unity when Q_i is $^n\Gamma_i$, are equal to the molecular weights when Q_i is $^m\Gamma_i$, and correspond to the molar volumes when Q_i is $^v\Gamma_i$. Combination of equations (2.51) and (2.52) gives

$$-\frac{d\gamma}{d\mu_2} = \Gamma_2 \left[1 + \frac{x_2 f_2}{x_1 f_1} \right] \tag{2.53}$$

and similarly it may be deduced that

$$-\frac{d\gamma}{d\mu_1} = \Gamma_1 \left[1 + \frac{x_1 f_1}{x_2 f_2} \right] . \tag{2.54}$$

Therefore the series of relationships may be written

$$-\frac{d\gamma}{d\mu_2} = {}^1\Gamma_2 \tag{2.55}$$

$$= {}^n\Gamma_2 \left[1 + \frac{x_2}{x_1} \right] \tag{2.56}$$

$$= {}^m\Gamma_2 \left[1 + \frac{x_2 M_2}{x_1 M_1} \right] \tag{2.57}$$

$$= {}^v\Gamma_2 \left[1 + \frac{x_2 V_2}{x_1 V_1} \right] \tag{2.58}$$

where M_i represents the molecular weights and V_i the molar volumes. A corresponding set of relations may be deduced for $-d\gamma/d\mu_1$.

2.9 CONCLUDING REMARKS
This chapter has concentrated on elucidating the effects of the presence of an interface on the systematic treatment of bulk phases with which we are familar.

Although the classical methods outlined in sections 2.4.1 and 2.4.3 have the dubious merit of apparent familiarity, the present authors are convinced that those employed in sections 2.4.2 and 2.4.4 have greater conceptual integrity.

The application of thermodynamic methods will be met again in later chapters, as and when considered appropriate. It should be remembered that the equilibrium thermodynamic considerations must apply in every equilibrium state that we choose to discuss. Statistical thermodynamic methods have much to contribute to the understanding of the interfaces, but their detailed consideration was taken to lie outside the scope of this book.

2.10 BIBLIOGRAPHY

J. G. Aston and J. J. Fritz, (1959) *Thermodynamics and Statistical Thermodynamics,* Wiley, New York, p. 241.

J. A. V. Butler, (1951) *Chemical Thermodynamics,* 4th ed., Macmillan, London, Ch. 21, p. 495.

J. W. Gibbs, (1928) *Collected Works,* 2nd ed., Longmans, New York, Vol. 1, p. 219.

E. A. Guggenheim, (1967) *Thermodynamics,* 5th ed., North Holland, Amsterdam, p. 47.

T. L. Hill, (1963/4) *Thermodynamics of Small Systems,* Benjamin, New York, Parts I and II.

J. G. Kirkwood and I. Oppenheim, (1961) *Chemical Thermodynamics,* McGraw-Hill, New York, Ch. 10, p. 148.

G. N. Lewis and M. Randall, (1961) *Thermodynamics,* 2nd ed., revised by K. S. Pitzer and L. Brewer, McGraw-Hill, New York, Ch. 29, p. 470.

CHAPTER 3

The study of liquid interfaces

3.1 INTRODUCTION

Liquid interfaces are usually the most easy to define, in contrast with solid surfaces which are the subject of the next chapter. The fundamental property of the liquid-vapour interface most susceptible to measurement is the *surface tension†*, which may be readily related to the interfacial energy in certain cases. The other properties of the liquid-vapour interface which are less frequently measured are the interfacial potential and interfacial viscosity, although most measurements are usually relative and interpretation sometimes difficult. The concentration of surface species adsorbed from a bulk solution has been measured directly by radio-tracer techniques in only a few cases, but frequently has to be deduced from other measurements.

Although planar liquid interfaces can be studied and are important, many practical situations involve interfaces that are curved. It is important to understand what effects such a situation has on such parameters as the vapour pressure above the liquid and the pressure differences across curved interfaces. These problems were tackled in the 19th century by Lord Kelvin and by Laplace, who derived theoretical equations which bear their names.

3.2 KELVIN'S EQUATION

This relation [1] describes the change in vapour pressure over an interface produced by variations in curvature. Let us consider for a moment the effect of atomising a quantity of bulk liquid. The interface would be greatly increased by such a process, and since the interfacial free energy is greater than the bulk free energy, work will have to be done on the system in order to carry out the process. In consequence the chemical potential of the material of the drops will be larger than that of the bulk liquid, and there will be a corresponding increase in vapour pressure over a convex liquid interface. The Kelvin relationship may be explicitly derived by thermodynamic reasoning.

---

†Although *interfacial tension* is the correct term, we shall in this book continue the use of this common description of the tension at the liquid-vapour interface.

In Chapter 2 we quoted the relation

$$dG = -SdT + VdP + \gamma dA + \Sigma_i \mu_i dn_i \tag{2.33}$$

which is applicable to a planar interface. However, it would remain true of curved interfaces so long as γ is unaffected by curvature. This equation gives rise to the definition of chemical potential

$$\mu_i = \left(\frac{\partial G}{\partial n_i}\right)_{n_j, T, P, A} . \tag{2.34}$$

Whilst this definition is useful in the description of the transport of material across a planar interface, for a small spherical droplet the addition of material to the drop must necessarily cause a change in A. The associated volume change could be written

$$dV = \Sigma_i v_i dn_i \tag{3.1}$$

where v_i is the partial molar volume of the ith component of the liquid. The volume and surface area of a drop are given by $4\pi r^3/3$ and $4\pi r^2$ respectively, thus $dV = 4\pi r^2 dr$ and $dA = 8\pi r dr$.
Therefore

$$dA = \frac{2dV}{r} = \Sigma_i \frac{2v_i}{r} \, dn_i . \tag{3.2}$$

If we combine equation (2.33) with (3.2) then

$$dG = -SdT + VdP + \Sigma_i \left[\frac{2v_i \gamma}{r} + \mu_i\right] dn_i . \tag{3.3}$$

The chemical potential μ'_i of the ith component in the drop is thus

$$\mu'_i = \left(\frac{\partial G}{\partial n_i}\right)_{n_j, T, P} = \frac{2v_i \gamma}{r} + \mu_i \tag{3.4}$$

or

$$\mu'_i - \mu_i = \frac{2v_i \gamma}{r} . \tag{3.5}$$

Now the chemical potential terms for the planar and curved interfaces may be written as

$$\mu_i = \mu_i^{\ominus} + RT \ln P_i$$

and

$$\mu'_i = \mu_i^{\ominus} + RT \ln P'_i$$

where P_i is the vapour pressure over a planar interface and P'_i that over a curved interface. If these relations are substituted in equation (3.5) then

$$\ln \frac{P'_i}{P_i} = \frac{2v_i\gamma}{rRT} \qquad (3.6)$$

which is a general statement of the Kelvin equation.

If instead of a drop of liquid in a vapour we were to consider a bubble of vapour in a liquid then the radius of curvature is assumed to be negative, and it is usual to write equation (3.6) in the form

$$\ln \frac{P'_i}{P_i} = -\frac{2v_i\gamma}{rRT} \qquad (3.7)$$

in these circumstances.

These last two equations have been derived for spherical interfaces, and the equation cannot be applied directly to other types of curvature. For an ellipsoidal surface, for example, equation (3.6) becomes

$$\ln \frac{P'_i}{P_i} = \frac{v_i\gamma}{RT} \left(\frac{1}{r_1} + \frac{1}{r_2} \right) \qquad (3.8)$$

where r_1 and r_2 are the radii of curvature.

The assumption we have used in this section so far is that the surface tension is independent of curvature, and for many systems with fairly large radii of curvature, say $r > 0.1$ μm, this is likely to be a reasonable assumption. However, as the radius approaches molecular dimensions this assumption is open to question as is also the assumption that it is realistic to think in terms of a radius of curvature. To illustrate this point, for a water drop of 1.0 nm radius the surface tension would fall from the planar interface value of 72 mN m^{-1} to about 54 mN m^{-1}, with only about 20 or so water molecules in the surface of the drop.

3.3 THE LAPLACE EQUATION [2]
As well as the vapour pressure being a function of curvature it is also interesting
to consider, for example, how the total pressure inside a bubble depends on the
radius of curvature. Let us assume that we have a spherical bubble of gas in
liquid (Fig. 3.1) in the absence of any external field (e.g. electrical or gravitational).

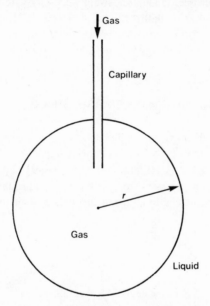

Fig. 3.1 — A bubble of gas in a liquid. The pressure of gas inside the bubble is P_g,
and the pressure in the liquid, P_l.

If we enlarge the bubble by introducing additional gas then the work done
can be expressed in terms of that done against the forces of surface tension and
that done in increasing the volume of the drop. Thus we could write

$$\gamma dA = (P_g - P_l)\, dV. \tag{3.9}$$

In deriving the Kelvin equation we deduced an expression for dA (equation
(3.2)), and if we also write $\Delta P = P_g - P_l$, then substitution into equation (3.9)
gives

$$\Delta P = \frac{2\gamma}{r} \tag{3.10}$$

which is the form of the Laplace equation for a spherical interface. This is the
specific form for a spherical meniscus of the more general equation for an
elliptical surface

$$\Delta P = \gamma \left(\frac{1}{r_1} + \frac{1}{r_2} \right)$$ (3.11)

where r_1 and r_2 are the principal radii. It is the fundamental equation governing the shape of all macroscopic menisci, and from it originate the more complicated treatments such as that of Bashforth and Adams [3]. This attempts to allow for the effect of gravitation on meniscus shape, and since normally surface tension measurements are made in gravitational fields, then such treatments have great utility.

3.4 CYLINDRICAL MENISCI AND THE MEASUREMENT OF SURFACE TENSION

Let us consider an example in relation to a situation which can be used, as will subsequently be described, to measure the surface tension of a liquid. Fig. 3.2 shows a plate suspended in a liquid and being maintained in its position by the force, f. The diagram therefore represents an equilibrium situation where this force is balanced by the combination of the weight of the plate and the meniscus force.

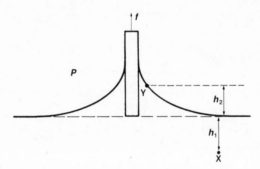

Fig. 3.2 – A plate suspended in a liquid.

In the present context of pressure variation across curved menisci in a gravitational field, let us look at the implications of equation (3.11) concerning meniscus shape. We are all familiar with the concept of additional pressure exerted upon a diver by the water above him. If in Fig. 3.2 the atmospheric pressure is P, then the pressure at a point X, a distance h_1 below the surface, where ρ is the density of the liquid, is

$$P_X = P + h_1 \rho g \quad .$$ (3.12)

Correspondingly at a point Y, above the general level of the surrounding liquid, the pressure will be given by

$$P_Y = P - h_2 \rho g \tag{3.13}$$

where h_2 is the distance above the general level of the surrounding liquid. Any point at the same horizontal level inside the meniscus will have the same pressure, thus in crossing the meniscus there will have to be a pressure increase from P_Y to P, and this would be given by the Laplace equation (3.11). Thus by combining equations (3.11) and (3.13) we obtain

$$h_2 \rho g = \gamma \left(\frac{1}{r_1} + \frac{1}{r_2} \right) \tag{3.14}$$

which predicts the variation of the elliptical radii term in such a way as to compensate for the variation in hydrostatic pressure as h_2 varies. However, it does not directly predict the complete shape of the meniscus, for example it does not set a maximum height of h_2, where the meniscus meets the plate, either tangentially or at some angle of contact. Lord Rayleigh [4] stated in his 1915 paper on capillary rise: "The case of a straight wall, making the problem two-dimensional, is easy . . . . . .", and he refers back to the appendix to an earlier paper (1892) of his concerning the reflectance of light from a mercury surface, in which [5] he was concerned with the problem of how far from the wall may the liquid surface be considered flat. In this he proposed two relations to describe the surface, which in the present notation are

$$x = \frac{1}{\sqrt{2}} \left(2a \cos \frac{\theta}{2} - a \ln \cot \frac{\theta}{4} \right) \tag{3.15}$$

$$z = \sqrt{2}\, a \sin \frac{\theta}{2}$$

where x is the horizontal co-ordinate and z the vertical. The capillary constant a is given by

$$a^2 = \frac{2\gamma}{\rho g} \tag{3.16}$$

and θ is the angle the tangent to the meniscus makes to the horizontal. At a large distance from the wall

$$\theta = 0, \quad z = 0, \quad x = \infty \quad ,$$

at the vertical edge of the wetted wall, $\theta = \pi/2$, and the origin of x corresponds to

$$\theta = \pi \quad \text{and} \quad z = \sqrt{2}a$$

which is not a particularly convenient definition.

A more useful way, in certain respects [6], of describing the meniscus is in the form of a single equation for the curve involving only the variables x and z, and the capillary constant,

$$x = (2a^2 - z^2)^{\frac{1}{2}} - \frac{a}{\sqrt{2}} \ln \left[\frac{z}{\sqrt{2}a - (2a^2 - z^2)^{\frac{1}{2}}} \right]. \qquad (3.17)$$

The universal curve predicted by this equation is shown in Fig. 3.3 where z/a is plotted as a function of x/a. The position where this meniscus meets the wall is determined by the contact angle, $(90-\theta)$ degrees. Thus if the wall is wetted and the contact angle is zero, then the maximum value of the ordinate would be 1.0 and the curve would be truncated at this point where $x/a = -0.3768$. This means that the meniscus has a clearly defined height, and that although the nature of the solid and its adhesive properties determine the contact angle, the shape of the meniscus away from the surface is not altered by the solid but only by the capillary constant.

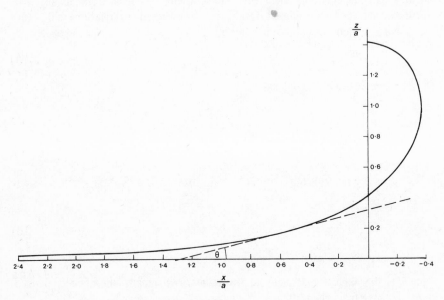

Fig. 3.3 — Profile of a cylindrical meniscus.

The volume of liquid supported by a unit length of a cylindrical meniscus may be found as follows:

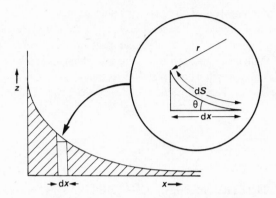

Fig. 3.4 – Cylindrical meniscus at a vertical plate.

Let us consider an element of the surface dS of radius r which supports a volume of liquid dV such that

$$dV = z \, dx \quad .$$ (3.18)

But using the Laplace equation

$$\gamma/r = \rho g \, z$$ (3.19)

and also

$$dx = dS \cos \theta = r \cos \theta \, d\theta$$ (3.20)

then

$$dV = \frac{\gamma \cos \theta \, d\theta}{\rho g} \quad .$$ (3.21)

Integration using the boundary condition that when θ is zero $\sin \theta$ and V are also zero, gives

$$V = \frac{a^2 \sin \theta}{2} \quad .$$ (3.22)

This is a form of the relationship used to calculate surface tension from Wilhelmy plate measurements [7].

The plate must be suspended exactly as shown in Fig. 3.2 otherwise there is a buoyancy correction. External pressure will result in a force applied to the plate unless the pressure at the top of the plate is equal to that at the bottom. If the bottom of the plate is below the general level then there will be an additional upwards force, f_u, given by the product of pressure and area, $f_u = \rho g h . b l$, where h is the distance of the bottom of the plate below the general level, and b and l the breadth and length of the plate respectively. This may also be written as $f_u = v \rho g$, where v is the volume of the plate below the general level of the liquid.

3.5 AXIALLY SYMMETRIC MENISCI AND THE MEASUREMENT OF SURFACE TENSION

The application of the Laplace equation (3.11) to other than the two-dimensional cylindrical meniscus is difficult. The solution of the equation is aided considerably if the meniscus shape possesses axial symmetry, for example a drop of liquid on a uniform clean horizontal surface. The more complicated case of, say, a drop on the curved surface of a tube, appears to have been largely ignored in the literature. Apart from the cylindrical meniscus method discussed in the previous section, all other common methods for measuring surface tension are based on axially symmetric menisci, and will be discussed in this section.

3.5.1 Capillary rise

There are a number of practically significant axially symmetric menisci which from a physical point of view have a lot of similarity, and each have been used to measure surface or interfacial tension. The simplest case is that of capillary rise in a cylindrical tube, which was considered by such notable early scientists as Poisson (1831) and Mathieu (1883). The experiment is comparatively easy to perform, provided that cleanliness is maintained. The capillary tubes must be cleaned thoroughly, and the method is usually only used if the tube is well wetted, that is the contact angle is zero. Although we shall first consider the rise in a single tube, in most experiments, because of the difficulty in accurately determining the general level of liquid, a differential method is employed, in which the difference in meniscus position between two tubes of differing internal radius is measured. The July 1915 paper of Richards and Coombs [8] prompted Lord Rayleigh to comment that it "reflects mildly upon the inadequate assistance afforded by mathematics", and in an almost immediate reply which appeared in the October 1915 issue of the *Proceedings of the Royal Society*, he proposed equations for narrow and wide tubes [4]. However, the first comprehensive treatment was essentially that [3] of Bashforth and Adams (1883), and was not considered by Rayleigh in his paper.

The basic geometry of a capillary rise experiment is shown in Fig. 3.5.

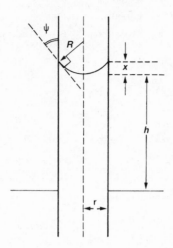

Fig. 3.5 – A liquid meniscus in a capillary tube.

In the simplest treatment, where the meniscus is assumed to be part of a sphere and the liquid above the level h (the height x) is ignored, then for a cylindrical tube we may write, using the Laplace equation (3.10),

$$\Delta P = h \rho g = \frac{2\gamma}{R} = \frac{2\gamma \cos \psi}{r} \tag{3.23}$$

or

$$\gamma = \frac{r h \rho g}{2 \cos \psi} \tag{3.24}$$

which, using the definition of the capillary constant (3.16), becomes

$$a^2 = \frac{r h}{\cos \psi} \qquad . \tag{3.25}$$

The form of this equation when the wall is wetted and the contact angle $\psi = 0$, was known before the 18th century, the first corrections being proposed by a London physician, Jurin [9], in 1718, when he allowed for the volume of liquid in a spherical meniscus by adding the term $r/3$ to the height, thus

$$a^2 = r(h + r/3) \tag{3.26}$$

for a wetted tube. Rayleigh's treatment for a narrow tube is a series extension of this equation correcting more exactly for deviation of the meniscus from spherical geometry:

$$a^2 = rh(1 + r/3h - \frac{2r^2}{3h^2}\ (\ln 2 - \tfrac{1}{2}) + \frac{r^3}{9h^3}\ (32 \ln 2 - 21))$$

$$= rh(1 + r/3h - 0.1288\ r^2/h^2 + 0.1312\ r^3/h^3) \qquad . \qquad (3.27)$$

This is very close to the result of Hagen and Desains [10] who assumed an elliptical form for the surface yielding

$$a^2 = rh(1 + r/3h - 0.1111\ r^2/h^2 + 0.0741\ r^3/h^3) \qquad . \qquad (3.28)$$

It is usually stated that these equations are adequate for tubes up to a diameter of 1 mm for water, which corresponds to a value r/h of approximately 0.02. Equations (3.27) and (3.28) can be thought of as the classical equation where the measured height, h, is multiplied by the series correction term in powers of r/h. These terms are comparatively small since at $r/h = 0.02$, the correction factor is 1.007, being made up as follows if equation (3.27) is used:

$$a^2 = rh(1 + 0.006\ 667 - 0.000\ 051\ 52 + 0.000\ 001\ 050) \qquad .$$

In other words if the Jurin equation (3.26) were to be used it would only be in error by +0.005%, and for the water example quoted above implies an accuracy in measuring r and h of the order of 1 μm. Likewise the correction factors predicted by equations (3.27) and (3.28) do not differ significantly, being 1.006 617 and 1.006 623 respectively.

Rayleigh also considered the case of very wide tubes for which

$$\frac{r\sqrt{2}}{a} - \ln\left(\frac{a}{h\sqrt{2}}\right) = 0.8381 + 0.2798\ \frac{a}{r\sqrt{2}} + \tfrac{1}{2}\ln\left(\frac{r\sqrt{2}}{a}\right) \qquad (3.29)$$

which can be used to show that in order to reduce h to the order of 1 μm for water, a 5 cm diameter tube would be required. The cases falling in between the equations listed above require the application of solutions first proposed by Bashforth and Adams [3].

The starting point is the combination of equations (3.11) and (3.12) to give

$$P_0 + \rho\ g\ z = \gamma(1/r_1 - 1/r_2) \qquad (3.30)$$

where z is the vertical height of the liquid measured from the apex of the meniscus where $r_1 = r_2$, and P_0 is the pressure at this point. This equation is then transformed into a non-dimensional form in which one parameter predicts the meniscus shape and the other its size.

Four menisci shapes are shown in Fig. 3.6, each being useful experimentally in determining surface tension. The shapes A and D correspond to the cases where the fluid inside the meniscus is lighter than that outside, and shapes B and C the converse. Let the two principle radii of curvature at S be R, in the plane of the paper, and $x/\sin \phi$ in a plane perpendicular to the plane of the paper, where ϕ is the angle between the tangent at S and the horizontal. The length PS corresponds to $x/\sin \phi$ and this radius rotates about the symmetrical axis OO'.

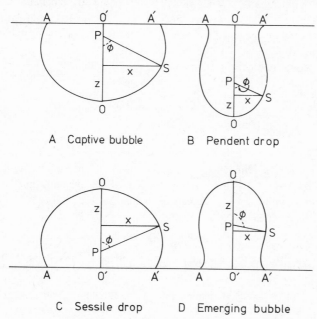

Fig. 3.6 – Shapes of axially symmetrical menisci.

At the apex of the menisci, O, where the two radii are equal, let them be b. At this reference point $z = 0$ and therefore from equation (3.30) the constant $P_0 = 2\gamma/b$. If ρ is the fluid density difference, then equation (3.30) becomes

$$2\gamma/b + \rho g z = \gamma(1/R + (\sin \phi)/x) \tag{3.31}$$

which may be rearranged to give

$$2 + \frac{\rho g}{\gamma} bz = \left[\frac{1}{R/b} + \frac{\sin \phi}{x/b} \right] . \tag{3.32}$$

If we make the further substitution

$$\beta = b^2 \rho \, g/\gamma \qquad (3.33)$$

it follows that

$$2 + \beta \, z/b = \left[\frac{1}{R/b} + \frac{\sin \phi}{x/b} \right] . \qquad (3.34)$$

The parameter β is that which defines the shape of the meniscus, and it has positive values for cases A and C, and negative values for profiles B and D of Fig. 3.6, where the relative density difference, ρ, possesses a negative value, being defined as the density inside the meniscus profile minus that outside it.

There is an alternative form of the Bashforth and Adams equation in terms of the volume contained within the meniscus, V, rather than R.

$$2 + \beta \, z/b = \left[\frac{\beta V}{\pi x^2 b} + \frac{2 \sin \phi}{x/b} \right] . \qquad (3.35)$$

In order to calculate the surface tension from capillary rise data it is sometimes necessary to use the extension of Bashworth and Adams' tables, first compiled by Sugden [11]. The relative accuracy of the various approximations used to interpret capillary rise data can be estimated from Table 3.1 — the last column gives the most exact result for values of $r/h > 0.03$.

Table 3.1

Comparison of the methods of calculating capillary rise results in the form

$$a^2 = rh \times F$$

r/h	Jurin term[8] (3.26) F	Rayleigh term[4] (3.27) F	Hagen term[10] (3.28) F	Sugden's Tables[11] F
0.000 01	1.000 0	1.000 0	1.000 0	1.000 0
0.000 1	1.000 0	1.000 0	1.000 0	1.000 1
0.001	1.000 3	1.000 3	1.000 3	1.000 3
0.01	1.003 3	1.003 3	1.003 3	1.003 2
0.03	1.010 0	1.009 9	1.009 9	1.010 0
0.06	1.020 0	1.019 6	1.019 6	1.019 6
0.10	1.033 3	1.032 2	1.032 3	1.032 1
0.15	1.050 0	1.047 5	1.047 8	1.047 2
0.20	1.066 7	1.062 6	1.062 8	1.062 2
0.50	1.166 7	1.151 0	1.148 2	1.144 4
0.70	1.233 3	1.215 6	1.204 3	1.194 2
1.00	1.333 3	1.336 7	1.296 3	1.263 6

Frequently the capillary rise method is used as a differential method, where the difference between meniscus heights in two different tubes, H, may be related to the surface tension by rewriting the usual form of the capillary constant equation $a^2 = bh$ as

$$H/(1/b_1 - 1/b_2) = a^2 \quad . \tag{3.36}$$

Since the actual radii of the tubes r_1 and r_2 usually differ appreciably, the whole of Sugden's tables are necessary for both b values to be estimated.

3.5.2 The captive bubble and sessile drop methods

The captive bubble and sessile drop geometries are illustrated in Fig. 3.6(A) and (C) respectively. In the first, as its name suggests, a bubble is blown below a flat horizontal surface, and in the second a drop of liquid is placed on the top of a flat horizontal solid surface. The surface or interfacial tension is determined from the shape of the drop or bubble. The actual experimental difficulties in determining the shape are considerable and similar to those of the pendent drop method, which will be discussed in the next section.

To obtain worthwhile results by either of these methods requires, after careful experimentation, lengthy, tedius calculation for an overall accuracy which is frequently worse than 0.1%. The calculation procedure is essentially one of fitting the experimental profile to a theoretical one derived from Bashforth and Adams' tables [3]. To make such comparisons all theoretical curves have to be scaled to a common maximum diameter, that of the drop, and the best fit value of β determined by the best apparent fit, after which the surface tension can be calculated from equation (3.33). A more detailed account may be found in the paper by Padday [6].

3.5.3 The pendent drop method

The geometrical parameters which have to be measured in order to calculate surface tension by this method are illustrated in Fig. 3.7.

A pendent drop elongates as it grows larger, because of the variation of hydrostatic pressure within it, causing variation from the value b at the apex in the curvature of the drop. This shape may be described by using equation (3.34), but it is difficult to measure β directly. However, as a shape determining parameter it may be related to other more easily measurable quantities. Andreas *et al* [12] suggested that this quantity should be $S = d_s/d_e$, where as shown in Fig. 3.7, d_s is the diameter measured at a distance equal to the equatorial diameter, d_e, up from the bottom of the drop. To overcome problems in determining b, the solutions to equation (3.34) were combined to produce a parameter $H = \beta(d_e/b)^2$, and tables are available (see Padday [6]) relating $1/H$ to S. Thus the surface tension may be calculated using

$$\gamma = \rho g b^2/\beta = \rho g d_e^2/\beta (d_e/b)^2 = \rho g d_e^2/H \quad . \tag{3.37}$$

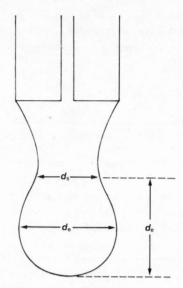

Fig. 3.7 — The pendent drop.

The optical problems [6] in measuring the drop with sufficient accuracy are severe, and a photographic image is usually used. Because of the small distances involved between the objective of the camera and the drop, the shape of the drop recorded on the camera film or plate is essentially that described by tangents from the optical centre of the lens to the drop, and hence allowance for this must be made. There are also problems in arranging for satisfactory illumination to avoid edge diffraction effects, and the fact that at the short distances involved camera lenses may show barrel or pincushion distortion. In fact a high quality enlarging lens may often prove to be better corrected for this type of distortion with the object and image distances normally encountered. The accuracy is again of the order 0.1% at best, but the calculations are easier than those for captive bubbles or sessile drops. The method is particularly useful for studying ageing surfaces, since the changes in the shape of the drop with time can be studied with a drop of constant volume, and the final shape is the true equilibrium shape yielding a value for the equilibrium surface tension.

An alternative method [13] of employing a pendent drop to measure surface and interfacial tensions is based upon the measurement of the height of the drop at the moment of instability, that is, when further increase in the drop volume results in detachment. No empirical corrections are required, but the limitation on accuracy is often that in actually measuring the height, and such results as have been presented in the literature suggest that the accuracy would normally be of the order of ± 0.10 mN m^{-1}. The authors introduce a characteristic length $l = (\gamma/\rho g)^{1/2}$, and two dimensionless variables $\lambda = r/l$ and $\kappa = h/l$, where r

is the radius of the tip from which the drop is hanging and h the height of the drop. The ratio of the experimental quantities r/h is equal to λ/κ. From tables of λ as a function of λ/κ a value of λ may be obtained by an interpolation procedure, and hence values of l and γ.

3.5.4 Maximum pull methods
Recently Padday *et al.* have proposed two new methods for measuring surface tension, either by the determination of the maximum pull exerted on a cylinder [14], or on a cone [15], after it is wetted and then slowly raised above the original liquid level.

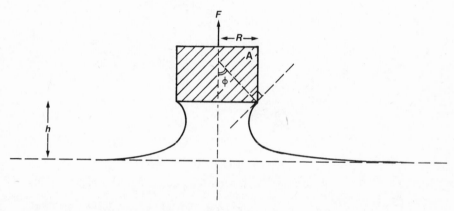

Fig. 3.8(a) – Geometrical relationships for the meniscus formed at the end of a cylinder.

The geometry for the cylinder case is shown in Fig. 3.8(a), where the broken vertical line represents the axis of the cylindrical rod A. It is assumed that the vertical wall remains unwetted. The force, F, in excess of the weight of the cylindrical rod, is exactly equal to the weight of liquid raised above the general level. Therefore

$$F = \pi R^2 h \, \rho g + 2\pi R \gamma \sin \phi \tag{3.38}$$

where R is the radius of the cylinder and h the height of the base of cylinder above the free surface. The first term represents the pressure on the underside of the cylinder and the second the resultant of the pull of the surface tension at the edge of the bottom of the cylinder. If F is divided by ρg the volume, V, of liquid raised will be found, thus equation (3.38) may be written as

$$\frac{V}{a^3} = \frac{\pi R^2 h}{a^3} + \frac{\pi R \sin \phi}{a} \tag{3.39}$$

using the present notation. At equilibrium V is related to R, h, a and through the Laplace equation[‡], thus

$$\frac{\mathrm{d}^2 z/\mathrm{d}x^2}{\sqrt{2}\,[1 + (\mathrm{d}z/\mathrm{d}x)^2]^{3/2}} + \frac{a \sin \phi}{2x} = z/a \quad . \tag{3.40}$$

This equation has been solved numerically by Padday [14].

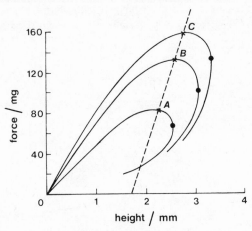

Fig. 3.8(b) − The force on a cylinder as a function of the height from a free liquid surface, for cylinders of various radii, $A = 0.141\,65$ cm, $B = 0.203\,92$ cm, $C = 0.239\,17$ cm, with water where $\gamma = 72.5$ mN m^{-1}, $l = 0.998$ g cm^{-3} and $g = 981.2$ cm s^{-2}. $\times$ = position of maximum force. $\bullet$ = position of maximum height. (After Padday [14]).

The force as a function of h generally takes the form shown in Fig. 3.8(b), and illustrates how for various cylinder diameters the force goes through well defined maxima. Padday has shown that the value of R/a may be expressed in terms of R^3/V, where V is calculated from the maximum pull $(V = F/\rho g)$, by a series of polynomials of the form

$$\frac{R}{a} = \frac{1}{\sqrt{2}}\left(A + B\left(\frac{R^3}{V}\right) + C\left(\frac{R^3}{V}\right)^2 + D\left(\frac{R^3}{V}\right)^3\right) \quad . \tag{3.41}$$

Practically the method is easy to use, requiring only a single-pan analytical balance with an accuracy of ±0.0001 g, a cylinder of appropriate size, a labjack and syringe (Fig. 3.9). Provided that the customary care over cleanliness can be maintained, then accuracies greater than ± 0.1 mN m^{-1} should be possible.

[‡] The expression from analytical geometry for the curvature of a line is $\dfrac{1}{R} = \dfrac{\mathrm{d}^2 z/\mathrm{d}x^2}{[1 + (\mathrm{d}z/\mathrm{d}x)^2]^{3/2}}$

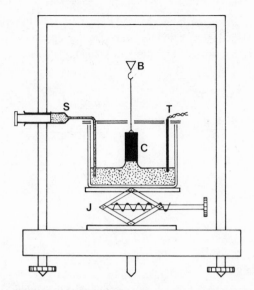

Fig. 3.9 — Apparatus for measuring the force on a cylinder using a single pan balance. S = syringe, B = balance, T = thermocouple, C = cylinder, J = lifting jack. (After Padday [15]).

The alternative method using a cone has been recently described [15] and is similar in certain respects. First of all the experimental method is essentially the same as that for the cylinder, and secondly the theory is based on similar principles. The geometry of the cone and meniscus is shown in Fig. 3.10. The analogous expression to equation (3.38) for the force in excess of the weight of the cone itself is

$$F = \pi R^2 H \rho g + 2\pi R \gamma \cos(\theta - \psi) - \pi R^2 \rho g (H + h)/3 \quad . \quad (3.42)$$

The first term of equation (3.42) represents the pressure acting at the plane of the meniscus where it meets the surface of the cone, the second term the vertical component of the pull due to the surface tension at the same location, and the third term the combination of the displacement of liquid due to the presence of the cone and the upthrust due to the liquid displaced by the cone. This may be rearranged to give the dimensionless form

$$\frac{V_{max}}{a^3} = \frac{2\pi R^2 H}{3a^3} + \frac{\pi R}{a} \cos \psi - \frac{\pi R^2 H}{3a^3} \quad . \quad (3.43)$$

Padday showed that from the maximum pull, the surface tension can be calculated from the cone constant, V_{max}/a^3, if the cone angle is known. The method is capable of accuracies greater than ± 0.1 mN m^{-1}.

Both of the last two methods can yield satisfactory values for single liquids and solutions provided the region around the maximum pull is investigated slowly in the solution case and reproducibility checked.

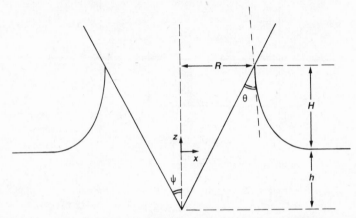

Fig. 3.10 – Cone with a liquid meniscus: geometrical relationships. The cone has a half-angle, ψ, and the liquid has a contact angle, θ, with the cone surface. The height, H, is that above the general level where the meniscus meets the cone, and h the distance below the general level of the cone apex.

3.5.5 Detachment methods

These methods include many of the most familiar surface and interfacial tension methods, namely the du Noüy ring and drop-volume methods, but they are frequently used for unsuitable systems. Whilst they yield good values for single liquids they are of limited use for solutions, because of the problems of surface ageing. During the measurements fresh surface is being continually created as detachment is approached, and errors of up to 25% have been reported.

The most well established is the du Noüy [16] ring method[†], although commercially made rings and balances are in fewer catalogues in recent years. The geometry of the experimental set-up is shown in Fig. 3.11.

The horizontal ring is dipped into the liquid surface and then slowly raised. The point of instability of the meniscus corresponds to the position of maximum force. Thus as the ring is slowly pulled out of the surface, the force exerted on the ring, and measured on the balance, increases steadily until the point is reached when the ring detaches itself from the liquid surface.

The simplest theory assumes that $R \gg r$, and that the inner and outer radii of the ring may be taken as equal; furthermore it is assumed that the inner and outer menisci approach verticality together and that this corresponds to the point of rupture. Thus the force applied at this time is given by

$$f = 4\pi R \gamma \quad . \tag{3.44}$$

[†]Padday [17] has traced earlier origins of the method.

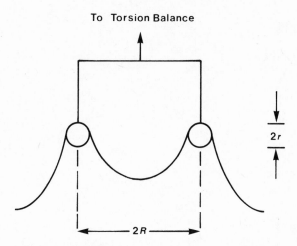

Fig. 3.11 – The du Noüy ring method: geometrical relationships. Not to scale.

There are a number of complications which need to be considered

1. The inner and outer radii are not equal.
2. As can be seen from the figure the inner and outer menisci do not approach verticality together. Thus at the rupture point neither surface is likely to be vertical and the radii of curvature are not likely to be equal to R.
3. When rupture occurs some liquid remains on the ring and can be seen there as small drops.

In an attempt to allow for these effects Harkins and Jordan [18] prepared tables of correction factors, F, as a function of the dimensionless parameters R/r and R^3/V, where V, the volume of liquid raised above the general level, is determined by $f = V\rho g$. The true surface tension is now calculated from

$$\gamma = \frac{f}{4\pi R} \ F(R/r, R^3/V) \ . \tag{3.45}$$

From a practical point of view the ring is first weighed in air and the additional weight measured at rupture converted to a force. Rupture is normally brought about by lowering the liquid surface slowly. It is important that the ring is clean, which can be achieved by flaming or immersion in chromic acid, and that it is hanging accurately horizontally. It has been stated that errors of $1°$ in inclination can produce errors of the order of 0.5%. Commercial rings are frequently made of platinum wire, and many have a value of R/r of 40. The correction factors applicable in this case are illustrated in Fig. 3.12. Accuracy can be of the order ± 0.1 mN m^{-1} for single liquids in many cases if care is taken, but with solutions errors up to 10% have been reported.

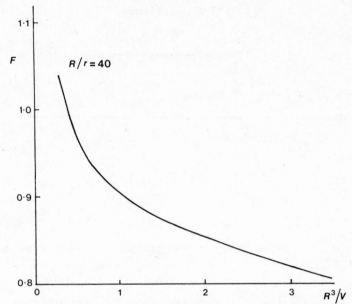

Fig. 3.12 – An example of the correction factors for use with the du Noüy ring for $R/r = 40$. (From W. D. Harkins and H. F. Jordan [18]).

The drop weight or drop volume method has again the virtue of reasonable accuracy for single liquids, but since the act of detaching a drop is controlled by gravity and cannot be slowed, errors can be expected with solutions. The process of detaching a drop from a tip is illustrated in Fig. 3.13. In fact the drops detach in pairs, with one drop many times larger than the other. Because of the extreme smallness of the drop forming from the elongating neck during detachment, it appears spherical in shape and is known as Plateau's spherule.

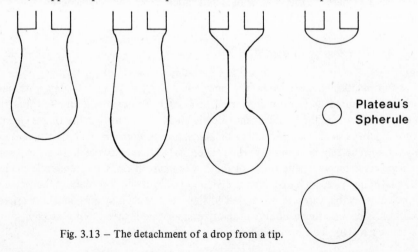

Fig. 3.13 – The detachment of a drop from a tip.

In practice most experimental procedures are constructed around a micrometer syringe (such as an 'Agla'). The syringe is fitted with an appropriate size tip, the criteria for which will be discussed shortly, and either the volume of the drop determined from the micrometer readings, or by collecting the drops and weighing them. In general it is normal for the tip end to be cylindrical, the drop forming on the end of the cylinder, which must therefore be well wetted. The junction between the end of the cylinder and its curved surface must be sharp and well defined, so that the radius of the tip may be measured with accuracy. If the drop volume or weight is to be measured with certainty, the gas phase in which the drop is formed will need to be pre-saturated with the vapour of the drop-liquid; or if interfacial tensions are to be measured the phases should be in equilibrium with each other.

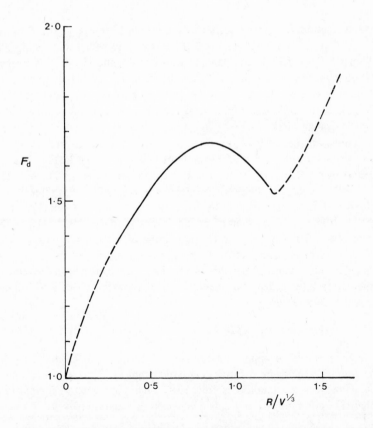

Fig. 3.14 – Drop-weight method correction factors, F_d of equation (3.47). The most accurate region is shown by a solid curve. (From Lando and Oakley [20].) Their factor $F_{LO} = 1/(2\pi F_d)$.

If the radius of the narrowest section at the moment of instability were to be equal to the radius of the tip, and the liquid precisely divided at this level, then under these conditions equating the gravitational pull to the restraining force of surface tension we have

$$Mg = V\rho g = 2\pi R\gamma \tag{3.46}$$

where M and V are the combined mass and volume respectively of both drops that fall, and R the radius of the tip. To overcome the obvious shortcomings of this simple theory, Harkins and Brown [19] proposed tables of correction factors, F_d, which are a function of the dimensionless parameter $R/V^{\frac{1}{3}}$. These are illustrated [20] in Fig. 3.14, and are used in the equation

$$\gamma = \frac{Mg}{2\pi R} \; F_d = \frac{V\rho g}{2\pi R} \; F_d \quad . \tag{3.47}$$

These correction factors are sufficiently accurate in the range $0.2 < R/V^{\frac{1}{3}} < 1.20$ to give surface tension values to ± 0.1 mN m^{-1} provided that the tip end from which the drop falls is circular and well wetted, and the drop is formed slowly. Consequently, in practice, a tip radius is chosen such that the experimental conditions lie within the above range. In practice the drop may be taken to about 95% of its final size rapidly and the final incremental volume added slowly over several minutes.

Some industrial standards require surface tension to be measured using this technique with a stalagmometer, an example of which is shown in Fig. 3.15. A bulb, B, of known volume, v_b, between marks E and F, has two graduated tubes attached of volume, v_d, per division. The stalagmometer is filled with liquid and then allowed to run out very slowly by controlling the closure of the tube at A, such that the drop rate is less than about one every four seconds. Immediately after a drop has fallen from the tip C, the reading of the upper meniscus D is taken, which is n divisions above E. Further drops are allowed to fall slowly from C and the total number, N, counted until the meniscus reaches a convenient level immediately after a drop has fallen, G, m divisions below F. The volume of one drop is then given by:

$$V = (v_b + (m + n)v_d)/N \quad . \tag{3.48}$$

It is perfectly possible to apply equation (3.47) correctly to calculate values of surface tension which can in favourable circumstances approach an accuracy of $\pm1.0\%$. However, the method is frequently recommended for comparitive measurements of two liquids 1 and 2, and hence from equation (3.47)

$$\frac{\gamma_1}{\gamma_2} = \frac{V_1\rho_1 F_{d_1}}{V_2\rho_2 F_{d_2}} \tag{3.49}$$

which is exact if the tip is wetted satisfactorily in each case. However, it is often stated that the two values of F_d are likely to be approximately the same and they are cancelled in equation (3.49). A glance at Fig. 3.14 should be sufficient to indicate that this is a dangerous procedure unless in both cases $R/V^{\frac{1}{3}}$ is close to 0.85. The objections to using this method for solutions have already been indicated.

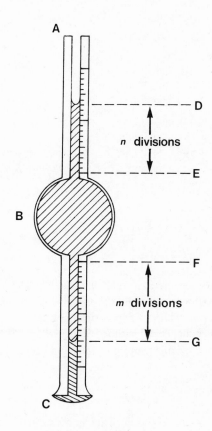

Fig. 3.15 – A design of stalagmometer.

3.5.6 The maximum bubble pressure method

This method again uses an axially symmetrical meniscus in an essentially unstable situation. The geometry of the bubble is shown in Fig. 3.16, and a suitable set-up for making measurements in Fig. 3.17. Once again the bubble profile may be described by the Bashforth and Adams approach [3]. The difference between the hydrostatic pressure outside the bubble, and the gas pressure inside, is attributable to surface tension forces and will be described by the Laplace equation.

Thus

$$P - \rho g t = \frac{2\gamma}{b} + \rho g z \quad .$$
(3.50)

If $P - \rho g t$ is replaced by its hydrostatic equivalent, h, then equation (3.50) can be rewritten as:

$$h = \frac{a^2}{b} + z$$
(3.51)

where as previously $a^2 = 2\gamma/\rho g$ and b is the radius at O. If also using Bashforth and Adams nomenclature $\beta = 2b^2/a^2$ and we put $X = a^2/h$ then equation (3.51) may be transformed to

$$\frac{r}{X} = \frac{r}{b} + \frac{r}{a} \cdot \frac{z}{b} \cdot \left(\frac{\beta}{2} \right)^{\frac{1}{2}}$$
(3.52)

where r/b is the x/b of Bashforth and Adams. The value of r/a is fixed in any measurement, although its value is initially unknown, and corresponding to it will be a series of values of r/X corresponding to a series of values of β and ϕ. For each of a sequence of assumed values of r/a, Sugden computed a series of values of r/b, and by substituting various β values in the relation

$$r/b = r/a \, (2/\beta)^{\frac{1}{2}} \quad ,$$
(3.53)

values of r/b may be obtained. Thus knowing r/b and β a value of z/b may be obtained from the tables and r/X calculated using equation (3.52). This value will be a maximum at the maximum bubble pressure condition. Sugden [21] tabulated his results as values of X/r for a series of values of r/a (reproduced in the Padday work [22]).

In practice to get reproducible results the radius of the origin has to be small, often less than 0.1 mm, and this places limitations on ultimate accuracy. However the method has been used to obtain surface tension values for molten metals, a problem for which the use of most of the other methods would prove difficult or impossible.

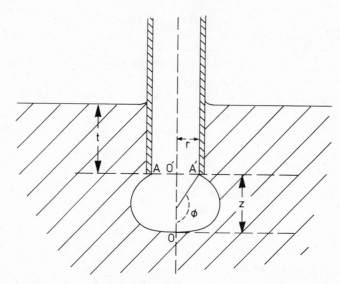

Fig. 3.16 – The maximum bubble pressure method: geometrical relationships.

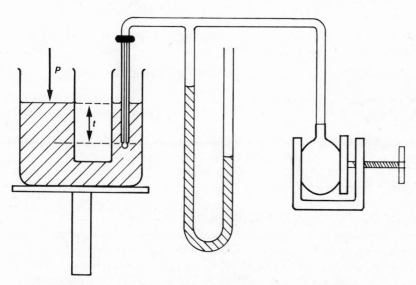

Fig. 3.17 – An experimental set-up for measuring surface tension by the maximum
bubble pressure method.

3.5.7 Comparison of methods for measuring surface tension

A comparison of the different methods for measuring the surface tension of
single liquids and solutions is given in Table 3.2.

Table 3.2
Methods of measuring the surface tension of single liquids and solutions

Method	Suitability	
	Pure liquids	Solutions
Capillary height	Very satisfactory when the capillary wets reproducibly.	Difficult when the contact angle is not 0° or variable.
Sessile drop	Very satisfactory.	Very useful for studying surface ageing.
Pendent drop	Very satisfactory but has experimental difficulties.	Useful for studying surface ageing.
Wilhelmy plate	With a good experimental set up, very accurate and convenient.	Provides accurate data on surface ageing.
Maximum pull on a cylinder	Very satisfactory. Easy to operate with simple apparatus.	Satisfactory if used with care and small displacements near maximum pull.
Maximum pull on a cone	Similar to cylinder method, but since the cone constant is universal, the results are easier to calculate.	
Du Noüy ring	Satisfactory	Unsatisfactory
Drop weight or drop volume	Very satisfactory	Poor when ageing effects suspected.
Maximum bubble pressure	Has experimental problems but useful where other methods are difficult to use.	Gives problems with ageing solutions.

3.6 THE MEASUREMENT OF INTERFACIAL TENSION

The measurement of the interfacial tension between two liquids presents a somewhat different set of experimental problems, and often the effects of contamination can be reduced, but should still not be ignored. It is well known that touching 100 cm^2 of clean water surface with one's finger can produce a surface tension lowering of 8 mN m^{-1} from the value of about 72 mN m^{-1} for a clean surface. The result on the interfacial tension of the benzene-water system, if benzene is added above the surface contaminated as just described, is not so readily measureable. On the other hand many of the methods we have considered for the measurement of surface tension, involve the formation of a stable meniscus at the junction between two liquid phases and solid. Although stable contact angles can be obtained for the liquid-vapour-solid junction, with two liquids the attractive forces between the respective liquids and the solid are often of similar magnitude, the contact angle is subject to hysteresis, and metastable menisci are common. As a result, usually the du Noüy ring, Wilhelmy plate and capillary rise methods are impracticable.

The drop weight or volume and maximum bubble pressure methods are not equilibrium methods, and variations of 1 mN m^{-1} have been reported for interfacial tension of two supposedly pure liquids. The maximum pull on a cylinder or cone does not appear to have been used so far for the measurement of interfacial tension, and they are only likely to prove satisfactory in particularly favourable wetting situations since they have the same set of problems as beset the du Noüy ring, Wilhelmy plate and capillary rise techniques.

The two methods most likely to yield consistent, reliable values for interfacial tension are those based on the profiles of pendent or sessile drops. It has been suggested that both of these methods are capable of good precision and good reproducibility. Furthermore, both methods are particularly suitable for measuring low interfacial tensions when the relative density is small, and measurements of values of $\gamma < 1$ mN m^{-1} have been attempted.

3.7 INTERFACIAL ENERGY AND INTERFACIAL TENSION OF SINGLE LIQUIDS

The surface tensions at room temperature of pure liquids in equilibrium with their vapours are usually within the range 10–80 mN m^{-1}, organic liquids being in the lower and water the upper part of this range. Interfacial tensions between hydrocarbons and water have values that lie between the surface tensions of the pure liquids. However, if the organic phase contains a polar group which can interact with the aqueous phase, then the resulting interfacial tension can be lower than either of the surface tensions of the two components. Some appropriate values are listed in Table 3.3, together with values of $d\gamma/dT$ and the work of adhesion (see below).

Fowkes' model [23] of the interface is illustrated in Fig. 3.18. Say, for example, liquid 1 is a hydrocarbon and liquid 2 is water, and 3 denotes the vapour phase. In the interfacial region of the hydrocarbon, the molecules are attracted towards the bulk hydrocarbon phase by dispersion forces which tend to produce a tension equal to the surface tension of the hydrocarbon, γ^{13}. However, at the interface there is also an attraction due to the dispersion forces between the hydrocarbon and water molecules in the interfacial region. This attraction can be predicted by the geometric mean of the dispersion force components of the surface tensions of hydrocarbons and water, $\sqrt{(^d\gamma^1 \ ^d\gamma^2)}$. Thus the tension in the interfacial region of the hydrocarbon is equal to $\gamma^{13} - \sqrt{(^d\gamma^1 \ ^d\gamma^2)}$. By a corresponding argument the tension in the interfacial region of the water phase will be $\gamma^{23} - \sqrt{(^d\gamma^1 \ ^d\gamma^2)}$. The measured interfacial tension, γ^{123}, will simply be the sum of these two tensions, thus

$$\gamma^{123} = \gamma^{13} + \gamma^{23} - 2\sqrt{(^d\gamma^1\ ^d\gamma^2)} \quad . \tag{3.54}$$

The work of adhesion between two phases, W_A^{123}, is defined as the work necessary to separate unit area of those surfaces (see p. 92) and is given by

$$W_A^{123} = \gamma^{13} + \gamma^{23} - \gamma^{12} \quad . \tag{3.55}$$

from which it can be seen that the quantity $2\sqrt{(^d\gamma^1\ ^d\gamma^2)}$ is the work of adhesion, when only dispersion forces are present. If other intermolecular forces are also present then usually $W_A^{123} > 2\sqrt{(^d\gamma^1\ ^d\gamma^2)}$.

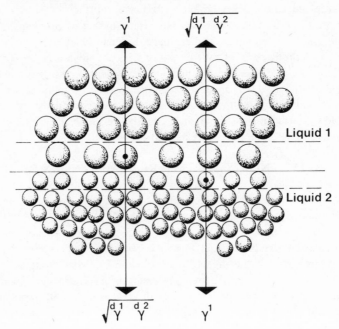

Fig. 3.18 – The Fowkes model of the liquid-liquid interface.

In Chapter 2 we have already considered the relationship between γ, A and G or G, but it is useful to look again at the equation

$$S\mathrm{d}T - V\mathrm{d}P + A\mathrm{d}\gamma + \Sigma_i\, n_i\mathrm{d}\mu_i = 0 \quad . \tag{2.45}$$

Now if we choose Gibbs' definition of a dividing surface and write S^s as the specific surface excess entropy at constant pressure, that is, the entropy per unit area of liquid surface less the entropy of the same amount of bulk liquid, then for a one-component system

$$\left(\frac{\mathrm{d}\gamma}{\mathrm{d}T}\right)_P = -S^s \quad . \tag{3.56}$$

The values listed in Table 3.3 show that this entropy term is positive, the explanation for which on a molecular level is not clearly understood. It would seem reasonable that molecules in a surface may be freer to move and more disordered than those in the bulk phase, but the appropriate model for the calculation of the configurational entropy is unclear.

Table 3.3

The surface and interfacial tensions of some liquids (γ/mN m^{-1}) at 293.2K together with values of W_A^{123}/mJ m^{-2} and (dγ/dT)/mN m^{-1} K^{-1}

	γ		W_A^{123}	$-$dγ/dT	
	Liquid/vapour	Water/liquid	Water/liquid	Liquid/vapour	Water/liquid
Water	72.75			0.16	
Octane	21.69	51.68	42.76	0.095	0.09
Dodecane	25.44	52.90	45.29	0.088	0.09
Hexadecane	27.46	53.77	46.44	0.085	
Benzene	28.88	35.0	66.6		
Tetrachloromethane	26.77	45.0	54.5		
Octanol	27.53	8.5	91.8		
Butanol	24.6	1.6	95.8	0.082	
Aminobenzene	42.9	5.9	109.8		
Diethylether	17.0	10.7	79.1		
Ethyl ethanoate	23.9	~3	~94		
Mercury	484	426	131		

Statistical thermodynamical methods have been used to calculate values of surface tension, entropy and energy, but have not given very accurate values. There are two approaches to the problem, the first based on the cell theory of liquids and the second on the radial distribution function method. The distribution of molecules around a given molecule does not average out to the bulk value near the surface because of relaxation, and the free volume of the liquid increases as the surface is approached. Using a spherical molecule model and pair-wise interactions calculated using say the Lennard-Jones 6-12 potential equation, the greatest density of nearest neighbours will be at the distance corresponding to the minimum in the calculated potential energy curve. The actual density of molecules as a function of distance is described by the radial distribution function. If this latter function can be determined experimentally by X-ray diffraction data, then it is possible to use the Lennard-Jones equation to calculate surface tension.

Using such a method, Kirkwood and Buff [24] obtained a value of 14.9 mN m^{-1} for the surface tension of liquid argon at 90K compared with an experimental value of 11.9 mN m^{-1}. The main problem with such an approach lies in the fact that X-ray data yield a bulk radial distribution function, and the experimental surface tension is the result of a relaxed surface state.

The cell model of liquids assumes that the bulk phase is divided up into cells with one molecule in a cell. Within a cell a molecule oscillates in the field generated by its nearest neighbours, and is further assumed to be independent of the motion of other molecules. The partition function, Q, is expressed as two terms, one for the molecules stationary in the potential minima at the centres of their cells, and the second for their kinetic and potential energy at other points within a cell. Using a square well potential model [25] and further assuming that the free volume of surface argon atoms is 30% larger than for bulk atoms, that is, some of the surface cells are empty, produced quite good agreement between theory and experiment.

The variation of surface tension with temperature may be related to the excess internal energy per unit area, U^s, by the Guggenheim-Goodrich [26, 27] equation in the form

$$U^s = \gamma - T\, d\gamma/dT = U - \Gamma' U_L \qquad (3.57)$$

where U is the internal energy of the molecules in the surface region, Γ' the total number of molecules in the surface region, and U_L the internal energy for a molecule in the bulk liquid phase. This measurement of $d\gamma/dT$ permits both S^s and U^s to be calculated.

There are several empirical equations which predict the variation of surface tension with temperature. The two most frequently used are the integrated form of the Eötvös [28] equation

$$\gamma(V_m)^{2/3} = k\,(T_c - T) \qquad (3.58)$$

where V_m is the molar volume, T_c the critical temperature and k a constant, which by comparison with the equation originally proposed by van der Waals [29]

$$\gamma = \gamma_0\,(1 - T/T_c)^{1.23} \qquad (3.59)$$

where γ_0 is a constant independent of temperature being the surface tension at 0 K, gives

$$k = \frac{\gamma_0\,(1 - T/T_c)^{0.23}}{T_c V_m^{2/3}} \quad . \qquad (3.60)$$

These equations work fairly well for non-polar liquids only, giving very poor results for water, for example.

3.8 POTENTIALS AT INTERFACES

The idea that interfaces have electrical properties is as old the real beginnings of the study of electricity itself. Galvani was an anatomist who in 1789 carried out some experiments upon the muscles of a frog's leg. He noticed a correlation between the muscles twitching and the simultaneous contact with both copper and iron. One metal was placed in contact with a nerve and the other with the muscle, resulting in muscle contraction. In Galvani's view this phenomenon was brought about by means of the metallic 'arc' of exterior negative charge being united with the positive electricity travelling along the nerve, the so-called 'animal electricity'. This theory led to a long and apparently friendly dispute with Volta, who showed that 'animal electricity' was no different from that associated with non-living materials. In a letter written in February 1794, some five years before he constructed the Voltaic pile, he attributed the contraction of the frog's muscle to the electrical differences of the two metals in contact with the tissue.

3.8.1 Galvani and Volta Potentials

Almost all interfaces can be, and usually are, associated with a potential difference. This can be the result of the distribution of elements of charge such as ions or electrons, or it may be due to the field resulting from permanent or induced dipoles. These electrical potential differences result in the redistribution of charges in the adjacent phase, and the formation of an electric double layer (see 3.8.3). There are a considerable number of early observations of phenomena associated with the presence of a double layer, for example the observation of electro-osmosis by Reuss in 1808, but the term was first used by Helmholtz in 1853. However, careful definition of the system and its elements is necessary if the potential between the interiors of two bulk phases with a common interface is to be understood. The approach outlined below has been described in detail by Parsons [30].

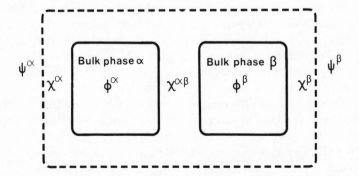

Fig. 3.19 — Diagrammatic representation of Galvani, Volta and chi-potentials in a two phase system (After Lange).

Consider two bulk phases, α and β, both of which contain the ionic species i (Fig. 3.19). When the system is other than at equilibrium, the electrochemical potential $\tilde{\mu}_i$ will differ in the two phases, and the work done in transferring an ion between the two phases, α to β, will be the difference in electrochemical potentials, $\tilde{\mu}_i^\beta - \tilde{\mu}_i^\alpha$. For convenience only one can consider this term as having two parts, the first due to the changing interactions between the ion and its differing surroundings in each bulk phase, $\mu_i^\beta - \mu_i^\alpha$, and the second concerned with any work done against the electrical potential difference, $\phi^\beta - \phi^\alpha$, between the two phases. The total work of transfer can then be written as

$$\tilde{\mu}_i^\beta - \tilde{\mu}_i^\alpha = (\mu_i^\beta - \mu_i^\alpha) + z_i e\,(\phi^\beta - \phi^\alpha) \tag{3.61}$$

which may be abbreviated for convenience to

$$\Delta^{\alpha\beta}\tilde{\mu}_i = \Delta^{\alpha\beta}\mu_i + z_i e\,\Delta^{\alpha\beta}\phi \tag{3.62}$$

where z_i is the valency of the ion and e the electronic charge. The transfer of uncharged species would mean that the work would be merely $\Delta^{\alpha\beta}\mu_i$. The electrical potentials in the interior of the bulk phase, ϕ^α and ϕ^β, are known as the **inner**-potentials, and $\Delta^{\alpha\beta}\phi$ as the **Galvani** potential difference. However, since $\Delta^{\alpha\beta}\tilde{\mu}_i$ cannot be measured by experimental electrochemical techniques, and $\Delta^{\alpha\beta}\mu_i$ is only determinable when $z_i = 0$, then in general $\Delta^{\alpha\beta}\phi$ is indeterminate.

The quantity $\Delta^{\alpha\beta}\phi$ is the change in electrical potential experienced by a probe passing from phase α to phase β. If one of the phases is a vacuum, then ϕ for the remaining phase, say ϕ^α, can be divided into two parts, the first of which ψ^α, is due to the electrostatic charge at the surface of the phase, and the second χ^α, is due to the presence of dipoles of any sort in the surface of the phase. Thus

$$\phi^\alpha = \psi^\alpha + \chi^\alpha \tag{3.63}$$

where ψ^α is known as the **outer**-potential of the phase. This name is chosen since although strictly the potential near a charged body in a vacuum varies continuously as the distance from the surface, electrostatic theory gives rise to a slightly unexpected conclusion. If we consider a sphere of 1 cm radius, then the potential is effectively constant in the region between about 10^{-4} to 10^{-2} mm, and most importantly equal to the average potential of the surface itself. As the sphere radius increases, the outer limit increases also, and the potential of this effectively constant region is ψ^α, and by using a probe in this region is accessible to measurement (see Fig. 3.20). By looking at equation (3.63) it is immediately obvious that the **chi**-potential, χ^α, is the potential difference between the interior of the bulk phase and the region where the potential is ψ^α, and also that χ^α will be positive when the potential increases from the outside to the inside of the phase. Neither ϕ^α or ψ^α are directly measurable.

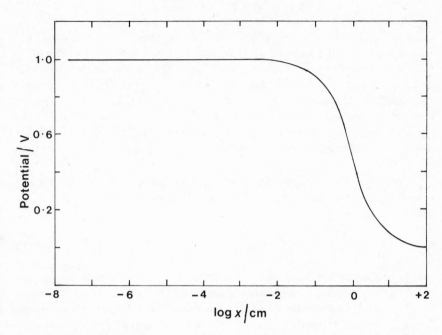

Fig. 3.20 – The potential near a sphere of 1 cm radius on which there is a net charge of 1.11 pC (After Parsons [30].

The electrochemical potential $\tilde{\mu}_i^\alpha$ may be written as

$$\tilde{\mu}_i^\alpha = \mu_i^\alpha + z_i e\ \phi^\alpha \tag{3.64}$$

and if phase α is a vacuum, then

$$\tilde{\mu}_i^\alpha = \mu_i^\alpha + z_i e\ (\psi^\alpha + \chi^\alpha) \tag{3.65}$$

which may be rearranged to give

$$\tilde{\mu}_i^\alpha - z_i e\ \psi^\alpha = \mu_i^\alpha + z_i e\ \chi^\alpha = \alpha_i^\alpha \tag{3.66}$$

where α_i^α is the **real**-potential of i in phase α.

When both α and β phases are present, that is, neither is a vacuum, then we may write

$$\phi^\beta - \phi^\alpha = (\psi^\beta - \psi^\alpha) + (\chi^\beta - \chi^\alpha) \tag{3.67}$$

and the quantity $(\psi^\beta - \psi^\alpha)$ is usually termed the **Volta** potential difference and can be measured. However, the terms on the r.h.s. of equation (3.67) assume measurements of phase α and β in a vacuum and not to a common interface between α and β.

Potentials at the interface between two bulk phases present a problem of some difficulty. The relative permittivity of the bulk phases will usually be greater than that of a vacuum and may also vary with distance from the interface, thus the significance of $\Delta^{\alpha\beta}\phi$ is less certain. The variation of potential with distance from the interface is more rapid, and it is not easy to define potentials equivalent to the outer potential. However, for a metal in a non-conducting dielectric, where apart from the influence of a constant relative permittivity the potential might vary as in a vacuum, then a term $\chi^{\alpha\beta}$ could be defined, but that due to molecular interactions across the interface it would not be equal to $(\chi^{\beta} - \chi^{\alpha})$.

The value of the chi-potential is essentially a function of the structure at a molecular level of the interface. Although χ cannot be measured, the Volta potential difference

$$\Delta^{\alpha\beta}\psi = \psi^{\beta} - \psi^{\alpha} \quad , \tag{3.68}$$

that is, the difference between the two outer potentials can be measured, and often changes in χ, for example when gaseous films are adsorbed onto interfaces.

The easiest example to use to illustrate this point is to consider two metals, α and β, in contact with each other but with no current flowing across the interface. We can then equate the electrochemical potentials of the electrons in each metal, thus

$$\widetilde{\mu}_e^{\alpha} = \widetilde{\mu}_e^{\beta} \tag{3.69}$$

from which we may deduce, using equations (3.63) and (3.64), that

$$\mu_e^{\alpha} - e\psi^{\alpha} - e\chi^{\alpha} = \mu_e^{\beta} - e\psi^{\beta} - e\chi^{\beta} \tag{3.70}$$

remembering that the electron is negatively charged, and on rearrangement

$$\Delta^{\alpha\beta}\psi = \psi^{\beta} - \psi^{\alpha} = (\mu_e^{\beta} - \mu_e^{\alpha})/e - \Delta^{\alpha\beta}\chi \quad . \tag{3.71}$$

If the value of $\Delta^{\alpha\beta}\psi$ is measured firstly for clean metal surfaces, $\Delta^{\alpha\beta}\psi_1$, and secondly for the case where an adsorbed film is at the interface, $\Delta^{\alpha\beta}\psi_2$, then

$$\Delta V = \Delta^{\alpha\beta}\psi_1 - \Delta^{\alpha\beta}\psi_2 = \{ [(\mu_e^{\beta})_2 - (\mu_e^{\beta})_1] - [(\mu_e^{\alpha})_2 - (\mu_e^{\alpha})_1] \}/e$$

$$- (\chi_2^{\beta} - \chi_1^{\beta}) + (\chi_2^{\alpha} - \chi_1^{\alpha}) \tag{3.72}$$

where ΔV is the Volta potential difference. Provided the adsorbed film is present solely at the interface and does not influence the bulk values of μ_e^{α} and μ_e^{β}, then the chemical potential differences on the right-hand side of equation (3.72) are zero. In certain circumstances further simplification is possible. At present we have that

$$\Delta V = (\chi_2^\alpha - \chi_1^\alpha) - (\chi_2^\beta - \chi_1^\beta) \tag{3.73}$$

but if we could arrange that adsorption takes place only in the α-phase, which is the experimental condition often attempted, then

$$\Delta V = \chi_2^\alpha - \chi_1^\alpha \quad . \tag{3.74}$$

The situation for a metal in an aqueous solution may be handled in an analogous way, except that the electrochemical potentials of the ions as well as the electrons need to be considered, but equations similar to (3.73) and (3.74) can still be deduced.

The most commonly investigated experimental situation is that of the effects of adsorbed films at the aqueous solution-air interface. Although the situation regarding insoluble monomolecular films spread at the interface may be handled by the last two equations, if a soluble film is present then the chemical potentials of the species in the bulk phase may be altered, the terms involving the differences in chemical potential of the species will no longer be zero, and more complicated relations of the type shown in equation (3.72) will apply.

3.8.2 THE EXPERIMENTAL MEASUREMENT OF VOLTA POTENTIAL DIFFERENCES

The basis of all the experimental methods, on various types of interface (metal-gas, liquid-vapour or liquid-liquid) is what is known as the **compensation method**, as is illustrated in Fig. 3.21. The method is based on the application of a potential opposite in sign to that associated with the interfaces between α and β, to produce a null reading on the current or voltage measuring device (detector).

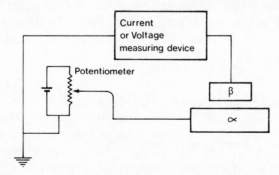

Fig. 3.21 – The measurement of compensation potential differences.

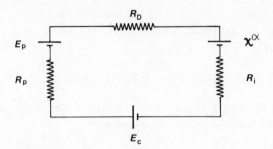

Fig. 3.22 – The equivalent circuit for the compensation method (simplified). R_i = the resistance of the gap between the measuring electrode and the conducting phase, χ^α = the chi-potential of the surface, R_D = the internal resistance of the detector, E_c = a compound potential of χ^β and all the other contact potentials in the circuit. E_p = the potentiometer potential, R_p = the potentiometer internal resistance.

The problem is that this gap between the phases can often be associated with a large impedance, and hence the detector is rendered insensitive to the balancing procedure. Let us imagine that for the moment we can express the equivalent circuit as shown in Fig. 3.22. Let us assume that we are concerned with changes at the α phase surface and that the surface of the β phase is unaltered by adsorption. In these circumstances we have the case where the resistance, R_i, of the gap between the α and β phases will be large. It will be dependent amongst other things on the distance separation and the humidity, if air is present in the gap: frequently it could be that $R_i > 10^{20}\ \Omega$. Normally one can arrange for $R_p < 10\ \Omega$, and therefore if the detector, having internal resistance R_D, is to measure accurately the difference potentials present, then $R_D \gg R_i$. This condition is impossible to meet experimentally, since electrometers do not usually have input impedances $> 10^{16}\ \Omega$, and often somewhat lower than this value. If, however, we could arrange for R_D and R_i to be not too dissimilar in magnitude, then although some sensitivity would be lost, the detector would record the potential difference multiplied by a factor. Therefore it is possible to accurately determine the potential difference when it is equal to zero, that is, when

$$E_p + E_c + \chi^\alpha = 0 \tag{3.75}$$

where E_p is potentiometer potential, which can be varied to suit the condition implied by equation (3.75), E_c is a compound potential of χ^β together with all the other contact potentials present due to the various junctions in the circuit, and χ^α is the chi-potential of the α-phase. Now if we imagine two conditions 1 and 2, the first with the surface of the α-phase clean, and the second in the presence of an adsorbed film, then we can write

$$E_{p_1} + E_c + \chi_1^\alpha = 0 \tag{3.76}$$

$$E_{p_2} + E_c + \chi_2^\alpha = 0 \quad . \tag{3.77}$$

On subtraction and using the definition of equation (3.74) then

$$\Delta V = \chi_2^\alpha - \chi_1^\alpha = E_{p2} - E_{p_1} \tag{3.78}$$

which is therefore determinable provided we can experimentally reduce R_i to the same magnitude as R_D. The only feasible experimental method of performing this task at present is to bombard the gap with α-particles from a radioactive source [31]. An example of a type of electrode suitable for this purpose is shown in Fig. 3.23. The original design incorporating $\sim$1 mC Po^{210}, a 5.3 MeV α-emitter, is no longer readily available, consequently the experimenter is now forced to consider the manufacture of his own device which can be based on Am^{241}, a 5.5 MeV α-emitter. Although the specific activities currently available are considerably less than for the original electrode, Am^{241} foil does have the advantages that larger areas can be easily ionised, and has a much longer half-life (458 years as opposed to 138.4 days). This would make an electrode designed around the foil a permanent piece of apparatus.

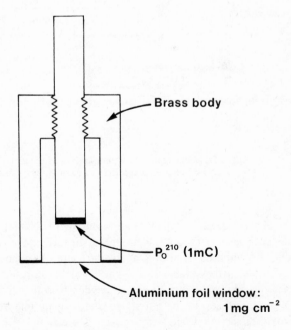

Fig. 3.23 – Diagrammatic representation of a Po^{210} electrode. Not to scale.

The alternative approach [32] to measuring the Volta potential difference is to make use of the high impedance of the gap between the α and β phases, rather than attempt to lower its impedance. If the electrode in the β phase is a gold plated disc some 2 cm in diameter, then this can be considered as forming a parallel plate capacitor with the surface whose capacity may be written as $C = A\, \epsilon_r \epsilon_o / d$, where A is the area of the plates and d their distance of separation. Thus altering d changes C, and we can consider, in view of our earlier remarks, that if the gap is small the small change would take place at constant potential, then since $Q = C V$, the change in C would result in the corresponding change in Q to cause a small current to flow in the circuit attached to the electrode. The best way to alter the gap width is to drive it sinusoidally, since this produces an alternating current in the measuring circuits, which is comparitively easy to amplify and measure. The compensation technique is still employed and a diagrammatic representation of such a circuit is shown in Fig. 3.24.

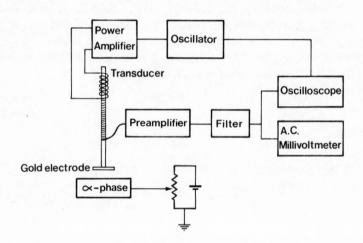

Fig. 3.24 – Diagrammatic representation of the vibrating plate method for measuring ΔV, as applied to a conducting α-phase-air interface.

In practice a frequency of a few hundred Hz is employed with the separation distance between the gold electrode and the surface of the α-phase of about 0.5 mm and the vibration amplitude about 0.01 mm. With modern equipment the maximum sensitivity obtainable is about 0.5 mV in favourable cases, and this method has the advantage that it can be applied to solid-gas, liquid-vapour and liquid-liquid interfaces. In the latter two cases usually one of the phases is an aqueous one, connection to it is by means of a calomel electrode, and the electrode vibrates in the non-conducting phase.

3.8.3 The interpretation of Volta potential difference measurements

The interpretation of ΔV measurements requires the consideration of molecular models of the interface. It is convenient to write

$$\chi^\alpha = \chi^\alpha \text{ (molecular dipoles)} + \chi^\alpha \text{ (ionic double layer)}, \qquad (3.79)$$

since in general both molecular dipoles and a distribution of charged ions could be expected at the interface. However, although the molecular dipole term is exactly described, this is not true of the ionic double layer. It is more precise to write

$$\Delta^{\alpha\beta}\phi = \Delta^{\alpha\beta}\phi \text{ (molecular dipoles)} + \Delta^{\alpha\beta}\phi \text{ (ionic double layers)}$$

$$(3.80)$$

(c.f. equation (3.72)).

Let us consider an array of molecular dipoles at a surface as a parallel plate capacitor with the distance of separation equal to the dipolar length, resolved in a direction normal to the surface. Then from electrostatic theory we may write

$$\chi^\alpha \text{ (molecular dipoles)} = n\ \mu_\perp/\epsilon_0 \qquad (3.81)$$

where n is the number of dipoles per unit surface area, $\mu_\perp$ is the component of each dipole moment perpendicular to the surface, and ϵ_0 the permittivity of a vacuum. The sign convention adopted requires that χ^α is positive if the positive ends of the dipoles are towards the bulk α-phase. If, and only if, the ionic double layer potential remains constant, then

$$\Delta V = \frac{\Delta(n\ \mu_\perp)}{\epsilon_0} \qquad (3.82)$$

and this equation has commonly been used to interpret ΔV measurements at air-water interfaces. However, the assumption that the ionic double layer potential can be maintained constant whilst the molecular dipole term varies is usually highly suspect, to say the least. There is no reason why they should be totally independent in general.

The theories governing the distribution of ions in a double layer have been frequently reproduced and reworked (see bibliography). The planar surface originally treated by Gouy and Chapman [33] may be used to illustrate the principles of the theory, and how the ionic double layer part of ΔV may be treated. The idealistic model they used as a starting point was that there exists a smeared out uniform surface charge, and that the ions in solution may be treated as point charges. At an infinite distance from the surface the electrical potential must be equal to the inner potential of the solution, but as the surface is approached the electrical potential gradually changes. If it is assumed that the work associated with bringing an ion from an infinite distance from the surface to some point nearer the surface is entirely electrical in origin, the distribution of ions in solution in the direction normal to surface can be described by the Boltzmann equation.

$$N_i(x) = N_i(\infty) \exp \{-z_i e \ (\phi(x) - \phi(\infty))/kT\} \quad . \tag{3.83}$$

Most treatments assume that all potentials are measured relative to the inner potential of the solution, and for convenience we shall adopt the same convention, thus equation (3.83) becomes

$$N_i(x) = N_i(\infty) \exp \{-z_i e \ \phi \ (x)/kT\} \quad . \tag{3.84}$$

In these equations $N_i(x)$ and $N_i(\infty)$ are the numbers of ions of type i per unit volume at distances x and ∞ from the surface, and z_i is the ion valency.

The relationship between the electrical potential and the net charge density per unit volume, $\rho(x)$, in relation to a planar double layer is given by Poisson's equation in the form

$$\frac{\partial^2 \phi(x)}{\partial x^2} = - \frac{\rho(x)}{\epsilon_r \epsilon_o} \tag{3.85}$$

where ϵ_r, the relative permittivity of the solution, is assumed invariant with x. But

$$\rho(x) = \Sigma_i z_i e \, N_i(x) \tag{3.86}$$

and thus combining equations (3.84), (3.85) and (3.86) gives

$$\frac{\partial^2 \phi(x)}{\partial x^2} = - \frac{1}{\epsilon_r \epsilon_o} \Sigma_i z_i e \, N_i(\infty) \exp \{-z_i e \ \phi(x)/kT\} \quad . \tag{3.87}$$

This may be integrated to give

$$\left(\frac{\partial \phi(x)}{\partial x} \right)^2 = \frac{2kT}{\epsilon_r \epsilon_o} \Sigma_i N_i(\infty) \{\exp (-z_i e \ \phi(x)/kT) - 1\} \quad . \tag{3.88}$$

The double layer as a whole, that is surface and solution, will be electrically neutral, therefore the charge per unit area of surface, σ, must be balanced by the charge in solution

$$\sigma = - \int_0^\infty \rho(x) \, dx \tag{3.89}$$

which when combined with equation (3.85) and integrated gives

$$\sigma = - \epsilon_r \epsilon_0 \left(\frac{\partial \phi(x)}{\partial x} \right)_{x=0} . \tag{3.90}$$

Substitution using equation (3.88) then gives

$$\sigma^2 = 2\epsilon_r \epsilon_0 \, kT \, \Sigma_i N_i(\infty) \left\{ \exp\left(-z_i e \, \phi(0)/kT \right) - 1 \right\} \tag{3.91}$$

where $\phi(0)$ is the potential at the surface, and will be equivalent to the ionic double layer contribution to ΔV. The artificiality of this separation of dipolar and ionic double layer effects cannot be over emphasized.

The more usually quoted form of equation (3.91) is

$$\sigma = (8N(\infty)\epsilon_r \epsilon_0 \, kT)^{\frac{1}{2}} \sinh \left(ze\phi(0)/2 \, kT \right) \tag{3.92}$$

for a single symmetrical electrolyte, that is, $N_+(\infty) = N_-(\infty) = N$ and $z_+ = z_- = z$. Integration of equation (3.88) after making similar simplifications gives

$$\kappa x = \ln \left[\frac{(\exp (u) + 1)(\exp (w) - 1)}{(\exp (u) - 1)(\exp (w) + 1)} \right] \tag{3.93}$$

where $u = ze\phi(x)/2 \, kT$ and $w = ze\phi(0)/2 \, kT$, and κ, known as the **reciprocal double layer thickness** is in general defined by

$$\kappa^2 = e^2 \Sigma_i N_i(\infty) z_i^2 / (\epsilon_r \epsilon_0 \, kT) \tag{3.94}$$

or in the symmetrical electrolyte case above

$$\kappa^2 = 2 \, e^2 z^2 N(\infty) / (\epsilon_r \epsilon_0 \, kT) . \tag{3.95}$$

Some values of κ and $1/\kappa$ are listed in Table 3.4.

Table 3.4

Values of κ and $1/\kappa$ for symmetrical electrolytes

C/mol dm^{-3}	κ/m^{-1} for 1–1 electrolyte	Values of $(1/\kappa)$/nm for various symmetrical electrolytes		
		1–1	2–2	3–3
1	3.256×10^9	0.31	0.15	0.10
10^{-1}	1.030×10^9	0.97	0.49	0.32
10^{-2}	3.256×10^8	3.07	1.54	1.02
10^{-3}	1.030×10^8	9.71	4.86	3.24
10^{-4}	3.256×10^7	30.71	15.35	10.24

Simpler forms of these equations may be derived when $ze\phi(x) \ll kT$, when it is legitimate to expand all exponentials and neglect all but the first and second order terms in the series. These give

$$\sigma = \epsilon_r\epsilon_0\kappa \; \phi(0) \tag{3.96}$$

and $\qquad \phi(x) = \phi(0) \exp(-\kappa x)$. $\qquad\qquad\qquad\qquad$ (3.97)

These last two equations will generally suffice when potentials are less than 25 mV in symmetrical electrolyte (1-1) solutions below 10^{-2} mol dm^{-3} at room temperature.

The interesting implications of both the exact equation (3.91) or its simplified form (3.96) is that if we assume constant $\phi(0)$, then the surface charge density, σ, becomes a function of electrolyte concentration, or conversely if constant charge density is assumed then $\phi(0)$ becomes a function of electrolyte concentration. The effect of electrolyte concentration for the constant potential and constant charge cases is shown in Fig. 3.25(a) and (b) respectively, and the variation in concentration of the relative types of ion in Fig. 3.26.

The limitation of Gouy-Chapman theory are primarily due to the inexactness of the initial assumptions. Firstly it is unlikely that the relative permittivity of a dipolar liquid such as water will be independent of electric field strength and the theory itself can predict fields strengths $\sim 10^8$V m^{-1}. There is a probability that this could result in reduction of ϵ_r by up to an order of magnitude. Secondly the point charge assumption for the ions in solution is likely to prove unsatisfactory for exactly similar reasons for the point molecule assumption failing to describe the behaviour of real gases, except when dilute. Ionic concentrations frequently exceed the ideal levels where Gouy-Chapman theory might work well, and also there is the complication noted by Stern [34] that the first layer of ions near

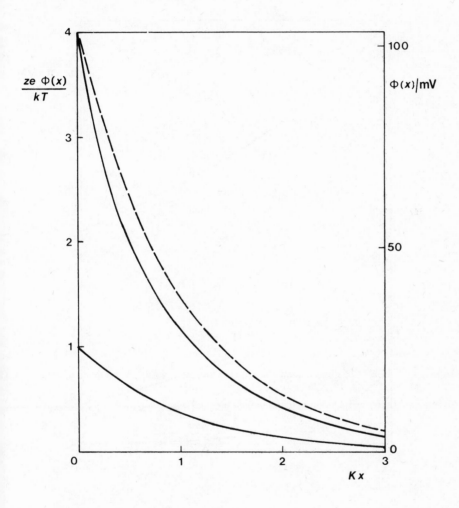

Fig. 3.25(a) – The constant potential case for the variation of the parameter $ze\phi(x)/kT$ as a function of κx. The full curve for $ze\phi(0) = 4$ is calculated using equation (3.93), and the dashed curve represents the approximate equation (3.97). When $ze\phi(0) = 1$, the values calculated from both equations cannot be separated on this graph, and are both represented by the full curve. A 1-1 electrolyte is assumed at 298.15 K.

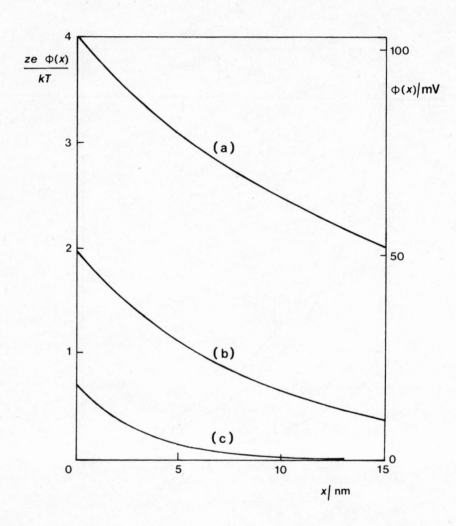

Fig. 3.25(b) – The constant charge case for the variation of the parameter $ze\phi(x)/kT$ as a function of x and the concentration of 1-1 electrolyte. The charge assumed is 4.30 mC m⁻² and the equation used is (3.93). $\epsilon_r = 80$ and $T = 298.15$K. The variation of $\phi(0)$ with concentration was calculated using equation (3.92). (a) 10^{-4} mol dm⁻³, (b) 10^{-3} mol dm⁻³, (c) 10^{-2} mol dm⁻³.

the surface will be limited in their distance of closest approach by their size; furthermore there will be chemical interactions involved. Thus the work involved in bringing an ion from the bulk of solution up to the interface will have both electrical and chemical components. However, the Gouy-Chapman theory can still be usefully employed to describe the region outside this first layer of ions adsorbed to the surface, where in the previous equations the quantity $\phi(0)$ is replaced by $\phi(\delta)$, the distance δ being the effective ionic radius of adsorbed ions. It should be remembered that ionic concentrations in the Stern plane will not be extrapolations of the curves shown in Fig. 3.26; such calculations as there have been indicate appreciable concentrations of co-ions as well as counter-ions.

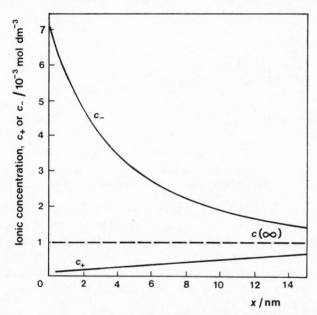

Fig. 3.26 – Graph of the concentration of univalent cations and anions as a function of distance from a positively charged surface ($\phi(0) = 50.4$ mV, $\sigma = 4.3$ mC m^{-2}, $\epsilon_r = 80$, $T = 298.15$K) where the bulk concentration is 10^{-3} mol dm^{-3}. Calculation using equations (3.93) and (3.84).

The third limitation is due to the assumption of a smeared out uniform surface charge, whereas it is known that it is in fact due to the net effect of discrete ions. In reality, therefore, the field near the surface is the result of overlapping ionic atmospheres from each individual surface ion. Thus at distances somewhat removed from the surface the result would be indistinguishable from the Gouy-Chapman assumption, but for the adsorption of ions in the first adsorption layer (Stern layer) the discreteness of the charges will affect individual ionic interactions. The potential distribution in the presence of a Stern layer is illustrated in Fig. 3.27.

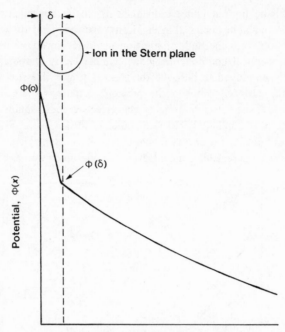

Fig. 3.27 – Diagrammatic representation of how the presence of a Stern Layer alters the distribution of potential near the surface.

To return to the problem of estimating the ionic double layer contribution to ΔV, we can now see that this is also composed of two parts. The magnitude of the contribution may be $\phi(0) - \phi(\infty)$, remembering that the inner potential may not necessarily be zero, but that the prediction of this contribution involves both Gouy-Chapman and Stern theories even in the most simple semi-realistic model. For further details the reader should consult the text by Sparnaay listed in the bibliography.

3.9 INTERFACIAL VISCOSITY
This subject is frequently under-represented in textbooks on surface chemistry (but not in Gaines [35]), and in certain respects this book is going to make the same error. This is largely because the results in the literature at present do not cover a sufficiently wide range of compounds adsorbed at interfaces to have established the techniques of measurement as routine. So for the present we shall confine ourselves to a discussion of fundamental definitions, the principles of measurement, and a few results which we consider significant. It seems likely that the capillary wave technique will ultimately reach the status of a standard method, since in principle satisfactory commercial apparatus could be readily available as soon as the demand appears and is recognised.

In order to approach the subject, it should perhaps be remembered that many measurements have been made on spread monolayers, where the results found were usually in the range $10^{-3} - 1$ 'surface poise'. This non-SI quantity has the units g s^{-1}, being the two-dimensional analogue of the three-dimensional bulk viscosity normally quoted in 'poise' corresponding to the units, g cm^{-1}s^{-1}. In the SI the dimensions of the bulk viscosity coefficient have become kg m^{-1}s^{-1} and the corresponding interfacial viscosity will have units kg s^{-1}. Thus we can define the interfacial viscosity η^s by

$$\text{Tangential force per m of surface} = \eta^s \times \text{rate of strain} . \qquad (3.98)$$

The relationship between the bulk viscosity, η, and the interfacial viscosity is

$$\eta = \eta^s/\tau \qquad (3.99)$$

where τ is the thickness of the interphase. In order to get some idea of the apparent bulk viscosity of the interphase from the values already quoted, say we put $\tau = 1$ nm for a spread monolayer then if we substitute in equation (3.99) we have $\eta = 10^3 - 10^6$ kg m^{-1}s^{-1}($10^4 - 10^7$ poise) assuming that the viscosity in this region is uniform. Davies and Rideal [36] likened this to suggesting that the interphase region would have the range of viscosities associated with butter.

The quantity defined above is the shear viscosity of the interface, but there also exists an interfacial dilational viscosity, κ^s, defined by

$$\Delta\gamma = \kappa^s \ \frac{1}{A} \left(\frac{dA}{dt}\right)_T . \qquad (3.100)$$

Thus κ^s is the fractional change in area per unit time per unit surface tension difference. From this quantity the analogous equilibrium property, the modulus of interfacial elasticity, E^s, can be defined

$$E^s = A(d\gamma/dA)_T = (d\gamma/d(\ln A))_T \qquad (3.101)$$

and the compressibility of the film

$$C^s = - \ \frac{1}{A} \left(\frac{dA}{d\gamma}\right)_T = \frac{1}{A} \left(\frac{dA}{d\Pi}\right)_T \qquad (3.102)$$

where Π is the surface pressure of the film defined by $\Pi = \gamma_0 - \gamma$, where γ_0 is the surface tension of the substrate without a film present and γ is the surface tension with a film present. Note that both E^s and C^s will be zero for a pure liquid since γ is independent of A. In certain circumstances with a surface active solute at high concentration a similar situation can be reached.

There are certain other analogies between the rheological properties of interphases and three-dimensional bulk phases which deserve consideration. Surface films which behave as simple liquids flow as soon as a shear stress (tangential force per unit length) is applied, and their flow properties may be characterised by η^s. These films may be Newtonian (η^s independent of shear rate) or non-Newtonian (η^s dependent on shear rate). However, more condensed surface films may behave in a fashion analogous to three-dimensional elastic solids, and need to be characterised by elastic moduli. We have already introduced the quantity E^s, which may also be called the interfacial compressional modulus, but there also is the parameter g^s, the interfacial shear modulus, which is defined as the ratio of the shear stress to the shear strain. If one wishes the analogies may be extended even further to the definition of an interfacial Poisson's ratio and interfacial Young's modulus [37].

The resistance of a liquid surface to deformation, as being something different from the bulk of the liquid, was first noticed by Plateau, who observed the differing rates of damping of an oscillating compass needle in the bulk of a liquid and in the surface. The majority of everyday liquids have surface-active materials present, thus the motion of a needle in the surface causes differences in surface tension, because of the non-equilibrium states created in front and behind the needle, thus dilational properties will evidently be present. Surface

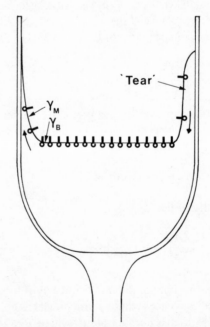

Fig. 3.28 — The Marangoni effect at the wall of a wine glass. Section chosen such that the liquid is rising on the left and falling on the right. Not to scale. Note the decrease in concentration of ethanol in the meniscus such that $\gamma_M > \gamma_B$.

tension derived forces generated in such a fashion, may also be generated by alternative means; for example the well known Marangoni effect, in which bulk liquid is transported as the result of surface tension gradients. Above the general level of wine in a wine glass (providing the wine is of sufficient strength!) 'tears' are observed on the glass and the wine appears to be in continuous motion (Fig. 3.28). The evaporation of ethanol from the meniscus film of wine on the glass leads to local increases in surface tension (from γ_B to γ_M) which in turn induces a surface and accompanying bulk wine flow upwards, the resulting accumulation of wine at higher levels returning as 'tears'.

The rate of movement of the surface is therefore determined by the surface tensional differences created by evaporation, and the interfacial dilational viscosity, and this surface motion drags some of the underlying bulk phase, probably down to depths of the order of 10^{-2} mm. If the reader likes to guess a few values for the imponderables, such as the surface evaporation rate, the diffusion rate of ethanol from the underlying bulk phase, the surface tension variation, etc., then calculations may be attempted and they suggest surface velocities of the order of cm s^{-1}.

3.9.1 The measurement of interfacial viscosity

The measurement of interfacial viscosity is not as easy as the average textbook might suggest. Certainly some of the apparatus looks remarkably simple, but obtaining reliable, reproducible and meaningful results is another matter altogether. Probably the most often used device used for measuring insoluble films of low Newtonian viscosity is the canal viscometer [38] (see Fig. 3.29) which is really a two-dimensional version of the Ostwald bulk liquid viscometer. The insoluble monolayer is permitted to flow from an area of higher surface pressure, Π_2, to one of lower surface pressure, Π_1, through a canal of length, l, and width, w, maintaining a constant difference in film pressure. In principle it should be possible to calculate the surface viscosity from an equation analogous to the Poiseuille equation for capillary flow, but correction has to be made for the drag on the surface film of the underlying bulk liquid. Harkins et al [39,40] derived a formula of the form

$$\eta^s = \frac{(\Pi_2 - \Pi_1)\, w^3}{12Ql} - \frac{w\, \eta}{\pi} \qquad (3.103)$$

where the second term is the bulk liquid drag term, and Q is area of monolayer flowing through the canal per sec, measured at the higher pressure, that is, $Q = (dA/dt)_{\Pi=\Pi_2}$. Certain approximations were made in the original derivation of equation (3.103), namely that the canal was narrow, $w \ll d$ (d = the depth of the canal) and long, $l \gg w$, that no slip occurs at the walls of the canal and that they are smooth, and that the film is flat, that is, the contact angle with the wall is 90°. Despite the problem of meeting these requirements the technique is still of use. It should be remembered that only an estimate of η^s can be made, and not the other rheological parameters. The value of η^s calculated represents some sort of average over a range of surface pressure $\Pi_2 \rightarrow \Pi_1$, and in practice, in order to make Q readily measurable, $\Pi_2 - \Pi_1$ has to be several mN m^{-1}. Furthermore, since the shear rate varies across the width of the slit, any non-Newtonian behaviour will also be averaged.

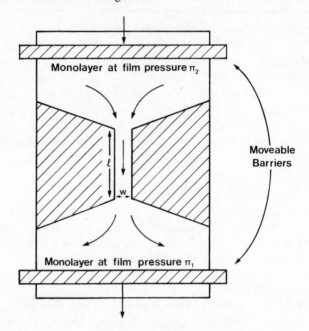

Fig. 3.29 – The principle of the canal viscometer (after Criddle and Meader [38]).

The second principle of measurement is that of the damping of the oscillation of a body moving in the surface. The apparatus is a simple torsion pendulum with a bob, suitable for the purpose, suspended so that it is touching the surface. The simplest forms are shown in Fig. 3.30. The platinum disk in the apparatus used by Langmuir and Schaefer [41] was 25.4 mm in diameter, and the moment of inertia of the bob was altered by changing the inertial rod. In the case of the knife-edged hollowed disk [38] the total inertial requirements may be met within the bob itself. Measurements are usually made relative to a clean liquid surface without an adsorbed film as follows.

The tortion pendulum is set in motion, and the torsional amplitude decreases with time. This is usually expressed in the theory derived by Tschoegl [42] in terms of the logarithmic decrement, λ,

$$\lambda = \ln (\theta_n/\theta_{n-1}) \qquad (3.104)$$

where θ_n is the amplitude of the nth swing. Because the changes between successive swings are small it is more accurate to use the form

$$\lambda = \frac{1}{n-m} \ln \frac{A_m - *A_m}{A_n - *A_n}$$ (3.105)

where A and $*A$ are the magnitudes of successive deflections either side of the rest point, and for the Newtonian behaviour n-m may be quite large. However, using successive amplitudes and plotting $\ln \theta$ vs. the number of swings to give the value of λ, has the advantage of showing up non-Newtonian behaviour. These measurements may be converted into viscosity parameters by the equations

$$\eta^s = \frac{I}{t_a} \left[\frac{1}{r_D^2} - \frac{1}{r_F^2} \right] \left[\frac{\lambda}{4\pi^2 + \lambda^2} - \frac{\lambda_0}{4\pi^2 + \lambda_0^2} \right]$$ (3.106)

$$g^s = \frac{I}{4\pi} \left[\frac{1}{r_D^2} - \frac{1}{r_F^2} \right] \left[\frac{4\pi^2 + \lambda^2}{t^2} - \frac{4\pi^2 + \lambda_0^2}{t_0^2} \right]$$ (3.107)

where t_a, t_0 and t are the complete periods of oscillation in air, touching the clean surface, and touching the film covered surface respectively. λ and λ_0 are the logarithmic decrements for the clean and film covered surface, I is the moment of inertia, and r_D and r_F are the radii of the disk or ring and film covered area respectively.

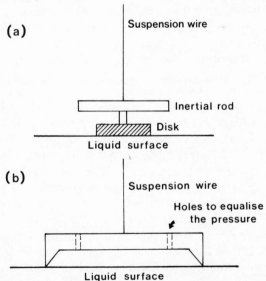

Fig. 3.30 – The torsion pendulum surface viscometer. Two alternative bob configurations: (a) the Langmuir and Schaefer design [41], and (b) the ring design of Criddle and Meader [38].

Very viscous films lead to problems as well. In this case the situation corresponds to overdamping, and the pendulum, after its initial displacement, returns slowly to its rest position, without oscillating. Langmuir [41] called this an 'aperiodic torsion pendulum', and it raises the problem that there is not a corresponding clean surface case. If δ is the angular displacement from the rest point, then a value of an apparent surface viscosity, η_a^s, with no sub-phase drag correction can be obtained from the relation

$$\eta_a^s = - \frac{\tau}{4\pi S} \left[\frac{1}{r_D^2} - \frac{1}{r_F^2} \right] \tag{3.108}$$

where S is the slope of the plot of $\ln \delta$ vs. time, and τ is the torsion constant defined by $\tau = 4\pi^2 I/t_a^2$.

The alternative way of using a torsion pendulum is that analogous to the Couette type of conventional rotational bulk viscometer [42]. In this case the liquid film is rotated at a particular speed, and δ is now the steady state displacement. Then for a constant angular velocity, ω, we have

$$\eta^s = \frac{\tau}{4\pi \omega} \left[\frac{1}{r_D^2} - \frac{1}{r_F^2} \right] (\delta - \delta_0) \tag{3.109}$$

where the subscript 0 denotes the clean surface condition as before. This method is really only useful for purely viscous films, since the constant shear rate condition can only be achieved after the elastic forces have been broken by exceeding the interfacial yield stress. Therefore clearly g^s cannot be measured. However, it is probably the best method for examining non-Newtonian behaviour of purely viscous films at the liquid-air interface.

3.9.2 Capillary waves, and their damping

It is an age old observation that small quantities of oil produce pronounced effects at the water-air interface. The English saying 'To pour oil on troubled waters' usually signifying a calming effect in an argument, clearly draws on this ancient knowledge. Benjamin Franklin[†] traces this observation back as far as Pliny (probably Gaius Plinius Secundus, 'Pliny the Elder', AD 23 or 24-79) and quotes other interesting examples, such as the calming of the surface of the sea by Bermudan fisherman using a little oil, so that the diffraction of light by little ripples is reduced, and fish in the water become easier to see and strike at when spear fishing. As a truly experimental scientist he attempted his own measurements, first on Clapham Common pond in London, and later, probably in 1772, on Derwentwater, a lake of some size in the Lake District of England, and again observed the calming of the water surface.

[†] See the June 2nd 1774 paper of Benjamin Franklin read to the Royal Society and reproduced in part in the *Appendix*.

There are three distinct physical processes associated with the presence of waves on a liquid surface; their formation, damping once formed, and the breaking of the larger waves. Using the example of oil on the sea, it is noticed that oil is largely effective in damping out the shorter wavelengths, which disturb the surface of larger waves encouraging the breaking of the latter. Numerous television pictures of oil tanker wrecks have shown this effect to the public too many times in recent years and no doubt will continue to do so. The oil also causes reduction of the wind drag on the larger waves.

The first laboratory experiments of this type were written up in 1891 by Fraulein Pockels [43]. This woman with modern ideas[t] showed that monolayers spread on the surface of water will dampen ripples on the surface. In fact this information could have been deduced directly from the data in Franklin's paper. He noted that 'a teaspoonful of oil' covered 'half an acre'. If we assume that the oil was probably olive oil and that the teaspoon contained 5 cm^3 (the value of the medical dose), then taking the 'half an acre' literally leads to film thickness of about 2.5 nm and molecular cross-section of about 0.21 nm^2. These values are in very close agreement with the formation of condensed monomolecular films of oleic (cis-9-octadecenoic) acid.

The velocity of wave propagation was considered by Sir W. Thomson in 1871 [44]. On the basis of simple physical principles he proposed that the velocity at the liquid-vapour interface of the wave, v, on the surface of a deep liquid corresponds to a wavelength λ, and are related by

$$v^2 = \frac{\lambda g}{2\pi} + \frac{2\pi\gamma}{\lambda\rho} \quad . \tag{3.110}$$

Since the first term is proportional to λ and the second to λ^{-1}, clearly at large wavelengths the first term will predominate and we have what are termed **gravity** waves, whereas at short wavelengths the second term will be much larger than the first and we then have what are called **capillary** waves. In the context of the present chapter only the latter are significant, and we can perhaps put some perspective on the situation from the following figures. If the surface wave has a frequency of 50 Hz on clean water at room temperature, then the wave velocity is about 30 cm s^{-1} with a wavelength of about 6 mm, and the capillary term of equation (3.110) is about 8.5 times larger than the gravity term.

For a pure liquid the surface tension is independent of surface area, and hence C^s is zero (see equation (3.102)). If we have an interfacial film present then the compressibility of the film will involve the expenditure of work during the extensions and contractions involved in the propagation of a ripple. In other

[t] She felt it unfair that her brother was allowed an academic career and that she should be condemned to the gentile life of her period. She apparently successfully persuaded her father to set up a laboratory for her in which she carried out her monolayer experiments (see C. H. Giles and S. D. Forrester, *Chem. Ind.* 43 (1971); 469 (1979)).

words the surface tension of the liquid is no longer a position-independent parameter, and may not be treated as a constant in the second term of equation (3.110). Now if the interfacial film is made up of soluble materials, say we are considering the surface of a sodium dodecyl sulphate (SDS) solution, then it is possible that adsorption-desorption phenomena will be involved in determining the variation of γ with the passage of a wave. Davies and Rideal [45] suggest that the relaxation time of the surface of 10^{-3} mol dm^{-3} SDS is of the order of 5 ms, and therefore at long wavelengths the surface tension would not vary significantly from its equilibrium value, whereas at very short wavelengths the interfacial film will behave more like an insoluble monolayer.

It would seem logical that fluctuations in surface concentration brought about by wave propagation will be influenced also by η^s, which can be thought of as determining the movement in the interfacial film of molecules in response to the surface tension variation. Dorrestein and Eisenmenger [46] have suggested that surface damping due to tensides (surface active agents) at frequencies in the MHz range may be largely due to the effects of interfacial viscosity. However, the theoretical models to describe such behaviour are still the subject of controversy, and attempts to use extended classical models to infer values of interfacial parameters may be subject to large errors. At the moment probably both improved models and better experimental data are required, but eventually it would seem that the study of the damping of capillary waves as a function of frequency could provide an extremely powerful tool for characterising the surface, both in respect of its visco-elastic properties and its adsorption-desorption equilibria.

3.10 SPREADING AND ADHESION IN LIQUID SYSTEMS

In 1869 Dupré formulated the concepts of the work of adhesion and the work of cohesion. Consider columns of an oil phase and a water phase in contact, with a common area of 1 m^2 (see Fig. 3.31). The work of adhesion, W_A^{OWV}, will be equivalent to the change in energy on separating the column,

$$W_A^{OWV} = \gamma^{OV} + \gamma^{WV} - \gamma^{OW} \tag{3.111}$$

and the corresponding definition for the work of cohesion of oil,

$$W_C^O = 2\gamma^{OV} . \tag{3.112}$$

These work terms are the result of intermolecular attractions across the interface. Thus if a particular oil-water intermolecular attraction is greater than either the oil-oil or water-water intermolecular attractions, then complete miscibility will occur, since the free energy will decrease on mixing. This situation can also be visualised in work terms. If $W_A^{OWV} > W_C^O$ and if $W_A^{OWV} > W_C^W$, then $W_A^{OWV} > \frac{1}{2}(W_C^O + W_C^W)$ and therefore equation (3.111) implies $\gamma^{OW} < 0$, the consequence of complete miscibility.

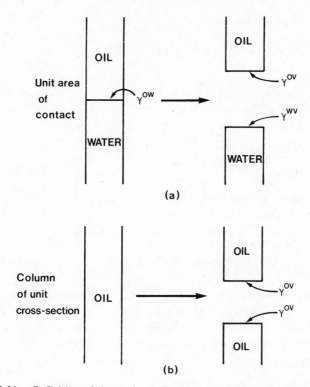

Fig. 3.31 – Definition of the work of adhesion (a) and the work of cohesion (b).

If we imagine a drop of oil remaining in an equilibrium position as a lens on the surface of water, then the geometric relationship between the interfacial tension forces will be as shown in Fig. 3.32.

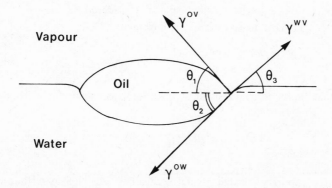

Fig. 3.32 – Diagrammatic representation of a drop of oil remaining as a lens on the surface of water.

If the forces are resolved horizontally then

$$\gamma^{WV} \cos \theta_3 = \gamma^{OV} \cos \theta_1 + \gamma^{OW} \cos \theta_2 \quad . \tag{3.113}$$

To achieve dynamic spreading of the oil film then the angles θ_1, θ_2 and θ_3 will approach zero and $\gamma^{WV} > \gamma^{OV} + \gamma^{OW}$. Harkins [47] used this to define an initial spreading coefficient, now called the spreading tension σ^{OWV}.

$$\sigma^{OWV} = \gamma^{WV} - (\gamma^{OV} + \gamma^{OW}) \quad . \tag{3.114}$$

One may also approach this from a thermodynamic point of view. If we assume that the drop increases its cross-sectional area in the plane of the water-vapour interface by dA, then if θ_1, θ_2 and θ_3 are small we can write

$$dA = (\gamma^{OV} + \gamma^{OW}) dA - \gamma^{WV} dA \tag{3.115}$$

using equation (2.10). Thus a positive spreading tension corresponds to an available energy decrease, and a spontaneous process.

Langmuir [48] showed that for larger drops the lens shape flattened, and that the maximum thickness of drop attainable, t_{max}, is described approximately by

$$t_{max}^2 = -2 \sigma^{OWV} \rho_w / g \rho_o (\rho_w - \rho_o) \tag{3.116}$$

where ρ_w and ρ_o refer to the densities of the water and oil phases respectively. This equation ignores the linear tension, f, (a one-dimensional analogue of pressure) which exists only along the three-phase junction at the perimeter of the drop perpendicular to the plane of Fig. 3.32. It is the force responsible for keeping the lens circular. The linear tension can be considered as the sum of three components, corresponding to each of the interfaces

$$f = f^{OV} + f^{WV} + f^{OW} \quad . \tag{3.116a}$$

Each term may be found by substituting the appropriate values in

$$f = \frac{2\sqrt{2}}{3} \gamma a [1 - \cos^3 (\theta_3/2)] \tag{3.116b}$$

where a is the capillary constant as defined earlier. For large paraffinic hydrocarbon lenses on water $f \sim 60$ μN, but of course since θ_3 will vary with drop size, so will f.

Using these ideas we can in a very simple-minded way attempt to understand the behaviour of a quantity of crude oil spilt at sea. Let us suppose we can treat crude oil as being composed of a light hydrocarbon fraction, a heavy hydrocarbon fraction, oxidised organic molecules (–OH, –COOH) and dissolved (or suspended) tar fractions. Harkins [47] quotes values of spreading tensions, listed in Table 3.5.

Table 3.5
Values of spreading tensions on water at 20°C

Substance	$\sigma^{owv}/mN\ m^{-1}$
1-octanol	+ 36.8
undecylenic acid (at 25°C)	+ 32.0
octane	+ 0.2
hexadecane	− 9.3

When the oil spill first hits the sea water, the surface active oxidised materials will concentrate at the interface and result in a large positive spreading tension, with resultant spreading of the slick. As the slick area increases two things will happen. First, the steadily increasing local area/volume ratio is likely to result in the surface concentration of the surface active oxidised materials being reduced, because of the finite amount within the slick, and further depleted by their solubility in the water. This will result in a progressive lowering of σ^{owv}, until it reaches $\sigma^{owv} < 0$, when lens formation is predicted. The maximum thickness will not be really described by equation (3.116), since it is extremely unlikely that true equilibrium exists, but the loss of the lighter hydrocarbon fraction by evaporation will result in an oily-tar of high viscosity being left on the surface of the water. The slick then appears to have definite physical dimensions, and on a relatively calm sea it would be reasonable to describe a slick as say '3 miles long and 1 mile across', moving in a particular direction.

One interesting example, which in less safety-conscious days used to make a nice lecture demonstration with a Petri dish on an overhead projector, is benzene on water. When a drop of benzene is initially placed on a clean water surface it spreads.

$$\sigma_i = 72.8 - (28.9 + 35.0) = 8.9\ mN\ m^{-1}\quad.$$

However, benzene and water are partially mutually soluble so that eventually

$$\sigma_f = 62.4 - (28.8 + 35.0) = -1.4\ mN\ m^{-1}$$

and the benzene will tend to retract to form lenses; σ_i and σ_f are the appropriate initial and final values of the spreading tension.

3.11 ADSORPTION PHENOMENA
3.11.1 Soluble films – Gibbs Monolayers

The experimental verification of the Gibbs adsorption isotherm (see equation (2.50)) is no longer a frequently reported experiment. The appropriate form of the equation for non-ionised materials has been fairly well tested, particularly by McBain *et al* in experiments which were both exotic and elegant. The situation for ionising materials is not quite so clearcut.

In order to use equation (2.50) we must substitute for the chemical potential

$$\mu = \mu^{\ominus} + kT \ln a \tag{3.117}$$

where $\mu^{\ominus}$ is the standard chemical potential and a the activity. Thus differentiating

$$d\mu = kT \, d(\ln a) = \frac{kT}{a} \, da \tag{3.118}$$

and if we write $a = fc$, where f is the activity coefficient, then

$$d\mu = kT \, d(\ln fc) = \frac{kT}{fc} \, d(fc) \quad . \tag{3.119}$$

For dilute solutions we may put $f = 1$, therefore

$$d\mu = kT \, d(\ln c) = \frac{kT}{c} \, (dc) \quad . \tag{3.120}$$

Substitution in equation (2.50) yields

$$-d\gamma = {}^{1}\Gamma_{2} \, kT \, d(\ln c) = {}^{1}\Gamma_{2} \, \frac{kT}{c} \, dc \tag{3.121}$$

and hence the value of ${}^{1}\Gamma_{2}$ may be predicted from measurements of the surface tension as a function of concentration.

One of the most quoted experiments in this context is that of McBain and Humphreys [49], normally called 'McBain's microtome'. It certainly represents a very ingenious experiment with most complex apparatus, which is illustrated in Fig. 3.33, and which McBain's original paper best describes:

> The trough B, Fig. 1, which contains the solution being used is made of pure silver 0.5 millimeters thick. It is approximately 7.5 centimeters wide, 85 centimeters long and the ends and sides are respectively three and eight millimeters high. It is supported between two heavy steel rails, R, as shown in Fig. 1. These rails are thirty-two feet long and it is upon them that the

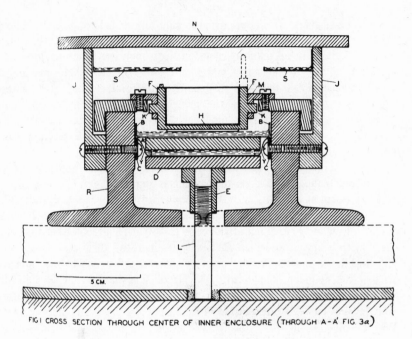

FIG.1 CROSS SECTION THROUGH CENTER OF INNER ENCLOSURE (THROUGH A-A' FIG. 3a)

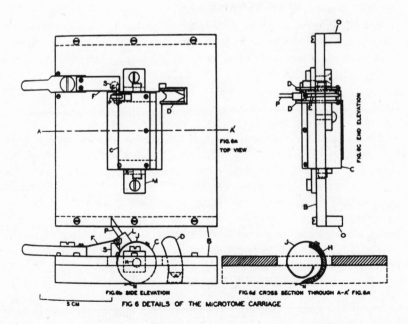

Fig. 3.33 – The microtome method of determination of the absolute amount of adsorption (Reproduced from McBain and Humphreys [49]).

microtome carriage, which supports the microtome blade, slides. The tracks are fastened securely to steel ties at intervals of two feet . . . . Near the outer edge of each tie is a set screw, C, which rests upon the concrete floor and allows adjusting of the height and levelling of the tracks. The machined upper surface of the rails was sufficiently true for the major part of the tracks. For the part adjoining the trough a much finer adjustment was necessary and was obtained by lapping this portion true so that the variation in level over the whole length of the trough was less than 0.015 mm. The upper part of both sides of the tracks are also machined.

The microtome carriage which slides along the tracks and carries the microtome blade which does the cutting is shown in detail in Fig. 6. It consists essentially of a frame, B, upon which is mounted the cylinder, C, which holds the blade. The cylinder consists of a section of one inch brass tubing, H, which is soldered at the ends to two brass disks. These disks are made with bearings, K, which fit closely through the angle pieces, M, thus allowing the cylinder to turn. The angle pieces are fastened to the carriage so that the cylinder is held firmly in place. A cover, J, made of one-half millimeter sheet silver is fastened over the remaining open part of the cylinder and curves around inside of it, thus acting as a baffle plate to keep the collected liquid from flowing out again. The microtome blade, N, is similar to a rigid safety razor blade. It fits into a small slot milled in the cylinder, as shown in Fig. 6d, and is soldered rigidly in place. The cylinder is lined with pure sheet silver and has a thin coating of paraffin so that the solution collected can be poured out more readily. The solution is removed by pouring it through one side of the same opening through which it enters. The cylinder is held in the proper position for cutting by the small steel pin, E, Fig. 6a, which rests upon the stop, S, and is held down by a thin strip of spring steel, F.

The microtome blade is given the speed necessary for cutting such a thin layer from the surface of the solution by shooting the carriage along the tracks by means of a slingshot arrangement made of rubber tubing. Since it is travelling quite rapidly (at the rate of 25 miles per hour) the carriage must be stopped rather rapidly after passing the trough. To retain the collected liquid the cylinder is turned through about 120° so that the blade and opening are pointing directly upward as soon as the trough is passed. It is turned by the steel pin, P, striking the stationary device, V, ... , and is held in this second position by a small braking device, D and E, Fig. 6c, which has a groove in which the steel pin, P, fits.

The microtome carriage is stopped quickly but without any sudden shock by the brake . . . . This consists of two steel strips, ten feet long, fastened to the outer sides of the rails. The braking action is exerted upon the two sides of the microtome carriage, O, Fig. 6c, which slide in the groove between the steel strips and the rails. The desired braking action is obtained by adjusting the pressure on the sides by turning the nuts on the ends of the bolts, B, which press upon the springs, F, and by regulating the width of the groove with the set screws, D.

There are certain comments that need to be made about the experiments before considering the results obtained. The relatively high speed chosen was presumably to assure that the surface layer removed is not significantly disturbed from equilibrium. The gravity wave travelling at the same speed as the microtome blade has a wavelength of about 70 m, thus no short wavelength ripples would be found in front of the blade.

The results were collected in a table and partially reproduced here as Table 3.6. At first sight the agreement between experimental and calculated values of the surface excess appears satisfactory. However, if the first series of 3-phenyl-propanoic acid (hydrocinnamic acid) data at a concentration of $1.5 \text{g kg}^{-1} H_2O$ is analysed (the only series presented in full in the paper), although the calculated surface excess is $5.1 \times 10^{-4} \text{g m}^{-2}$ and the mean experimental value $5.6 \times 10^{-4} \text{g m}^{-2}$, the standard deviation of the experimental results is $1.7 \times 10^{-4} \text{ g m}^{-2}$. If this is typical, then the apparent agreements are not the conclusive, but to quote McBain,

> These results represent the only existing measurements of the absolute amount of adsorption at *static* air-water interfaces. It can be seen that, under the best conditions, the observed values have agreed quite closely with those predicted by the Gibbs equation . . . ,

which was fair comment in 1931 when he wrote the paper.

Table 3.6
McBain's microtome experimental results [49]

Substance	No. of expts.	Concentration/ mol dm^{-3}	Surface excess/μmol m^{-2}	
			exptl.	calc.
4-amino-toluene	11	0.018 7	5.7	4.9
	29	0.016 4	4.3	4.6
phenol	18	0.218	4.4	5.1
hexanoic acid	30	0.022 3	5.9	5.4
	14	0.025 8	4.4	5.6
	43	0.045 2	5.3	5.4
3-phenyl-propanoic	33	0.010 0	3.7	3.4
acid	19	0.030 0	3.6	5.3

If McBain's experimental ingenuity with his 'railroad apparatus' was not as successful as he might have wished, he refused to be beaten. It was the constant realignment of the rails that finally lead to attempting the problem of analysing deeper layers of liquid to determine the surface excess. This surface compression method was just as ingenious but much simpler [50]. The diagrammatic representation of the apparatus in Fig. 3.34 illustrates how it works.

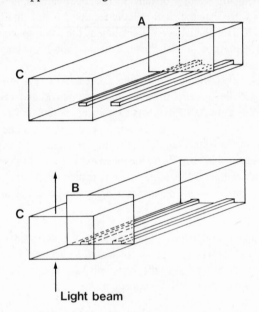

Fig. 3.34 – McBain's surface compression method. Position A before compression, position B after.

With a moveable barrier in position A, the surface between A and the end of the trough C is allowed to come to thermodynamic equilibrium. The barrier is moved rapidly from position A to position B, the aqueous sub-phase flowing backwards under the barrier between the skids. When the movable barrier reaches B it falls to the bottom of the trough, closing off the section BC from the remainder of the trough. The original surface between A and C is now compressed to the area between B and C, and to re-establish thermodynamic equilibrium some of the interfacial material redissolves, the total concentration being estimated interferometrically. The results he obtained with 0.5 g dm^{-3} solutions of dodecyl-sulphonic acid were somewhat better. The Gibbs equation gave a surface concentration of 1.44×10^{18} molec m^{-2} which agrees quite well with the experimental value $(1.57 \pm 0.15) \times 10^{18}$ molec m^{-2}.

The use of radiotracer counting techniques should offer an accurate and illuminating method of studying surface adsorption, and hence the opportunity of verifying Gibbs equation. However, it would seem that very little work has been carried out using non-ionising surface active materials, and some of that using ionising tensides contains experimental blunders.

For an ionising tenside the form of Gibbs equation is less certain. If we consider SDS, for example, since it is a strong electrolyte it can be considered to dissociate completely

$$C_{12}H_{25}SO_4Na \rightarrow C_{12}H_{25}SO_4^- + Na^+$$
$$\text{SDS} \qquad\qquad \text{DS}^- \qquad \text{S}^+ \quad.$$

The appropriate form of the Gibbs equation (2.48) will be

$$-d\gamma = {}^1\Gamma_{DS^-}\, d\mu_{DS^-} + {}^1\Gamma_{S^+}\, d\mu_{S^+} \quad. \tag{3.122}$$

If we expand the chemical potential terms in analogous way to equations (3.117) to (3.120), then we obtain

$$-d\gamma = kT\,[{}^1\Gamma_{DS^-}\, d(\ln c_{DS^-}) + {}^1\Gamma_{S^+}\, d(\ln c_{S^+})]$$

$$= kT\left[{}^1\Gamma_{DS^-}\,\frac{dc_{DS^-}}{c_{DS^-}} + {}^1\Gamma_{S^+}\,\frac{dc_{S^+}}{dc_{S^+}}\right] \quad. \tag{3.123}$$

In order to simplify this equation we must assume electrical neutrality is maintained in the interface, then we may write

$$^1\Gamma_{SDS} = {}^1\Gamma_{DS^-} = {}^1\Gamma_{S^+} \tag{3.124}$$

and $\quad c_{SDS} = c_{DS^-} = c_{S^+}$

which on substitution in equation (3.123) gives

$$-d\gamma = 2kT\,{}^1\Gamma_{SDS}\, d(\ln c_{SDS}) = \frac{2kT}{c_{SDS}}\,{}^1\Gamma_{SDS}\, dc_{SDS} \quad. \tag{3.125}$$

By comparing equation (3.121) with equation (3.125), it will be seen that they differ by a factor of 2, and that the appropriate form will need to be used in any experimental test of Gibbs equation. It is also quite clear that any partial degree of ionisation would lead to considerable difficulty in applying the Gibbs equation. Finally whilst on this topic, if SDS were investigated in a solution containing a large excess of sodium ions, produced by the addition say of sodium chloride, then the sodium ion term in equation (3.123) will vanish, and we will arrive back at an equation equivalent to (3.121).

Salley *et al* [51] reported in 1950 the results of their studies on S^{35} labelled di-octyl-sulphosuccinate. The choice of S^{35} was so that, like the other soft-β emitter C^{14}, the range of the β particles from surface active radio-labelled molecules in the aqueous phase would be very short, and that a Geiger tube with a relatively thin end window would count primarily the activity at the liquid-air interface.

In principle the count with S^{35} labelled di-octyl sulphosuccinate could be corrected for S^{35} in the bulk by measuring S^{35} labelled non-surface active sulphate, such as sodium sulphate, and in this way an accurate estimate of the surface excess obtained. The accuracy of their comparison, however, was ruined by their choice of the du Noüy ring method to measure surface tension and to evaluate from these results surface excess values using Gibbs' equation. They claimed a linear relationship between γ and ln c over the concentration range 2.3×10^{-6} to about 6×10^{-4} mol dm^{-3}, corresponding to a surface excess of 2.4×10^{14} molec m^{-2}, if equation (3.121) was used as a basis whereas the expected value should be given by equation (3.125), that is, 1.2×10^{14} molec m^{-2}. The counting technique showed a normal shaped adsorption isotherm with no plateau, and values ranging up to about 3×10^{14} molec m^{-2}. As a result of discussion with Onsager it was proposed that the equilibria present resulted in the undissociated acid being surface active species.

$$
\begin{array}{ccc}
\text{C}_8\text{H}_{17}\text{OOC CH}_2 & & \text{C}_8\text{H}_{17}\text{OOC CH}_2 \\
| & \rightleftharpoons & | \quad\quad +\text{Na}^+ \\
\text{C}_8\text{H}_{17}\text{OOC CH SO}_3\text{ Na} & & \text{C}_8\text{H}_{17}\text{OOC CH SO}_3^- \\
& & + \\
& & \text{H}^+ \\
& & \updownarrow \\
& & \text{C}_8\text{H}_{17}\text{OOC CH}_2 \\
& & | \\
& & \text{C}_8\text{H}_{17}\text{OOC CH SO}_3\text{H}
\end{array}
$$

In this case the appropriate form of the Gibbs equation would be

$$^1\Gamma_2 = -\frac{1}{RT}\frac{d\gamma}{d(\ln c)} \tag{3.121}$$

an unexpected yet interesting result.

The most convincing experimental proof of the Gibbs equation are the experiments of Tajima *et al* [52] who used tritium labelled SDS and made measurements at the solution-air interface. Their results shown in Fig. 3.35 show excellent agreement between equation (3.125) and experiment, provided that an appropriate surface tension measurement method is used, indicating that the salt in this case was the surface active species.

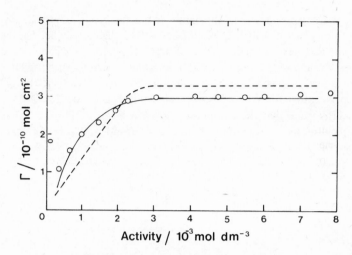

Fig. 3.35 – Comparison of observed and calculated values of SDS adsorption.
–o– observed values. The two lines represent calculations from surface tension
data derived ---- by the drop volume method, and —— by the Wilhelmy plate
method. (From Tajima *et al* [52]).

The point that was made earlier in the chapter, about the unsuitability of
non-equilibrium methods for measuring the interfacial tensions of tenside
solutions, where typically ageing effects are present, can be clearly seen. The
Wilhelmy plate method results do seem to be reliable. The very slight deviations
present are probably due to no allowance being made for the presence of H^+ ions
in the Gibbs equation calculations. The magnitude of the errors involved in
drop-volume surface tension measurement can be seen in Fig. 3.36(b), where
comparison is made with the Wilhelmy plate method. If the substrate also
contains 0.1 mol dm^{-3} sodium chloride as a swamping electrolyte, the form of
the Gibbs equation reverts to equation (3.121) since the concentration of Na^+
can be considered invariant and hence $d\mu_{s^+} = 0$, as previously discussed.

Taken in isolation this result might seem to raise questions over the conclusions of Salley *et al* [51] concerning di-octyl sulphosuccinate, but sodium
octadecanoate seems to behave similarly [53]. It is clear that the various possible
equilibria present can lead to somewhat unexpected conclusions concerning the
major surface active species present at the solution-air interface.

Fig. 3.36 illustrates clearly the relationship between the slope of the γ vs ln c
plot and surface concentration for SDS. The form of the Gibbs equation (3.125)
indicates that if Γ is to be constant then $d\gamma/d(\ln c)$ must also be constant. Thus
from Fig. 3.36(a) one would predict that $d\gamma/d(\ln c)$ would be constant from a
concentration of about 3×10^{-3} mol dm^{-3} upwards, and comparison with
the Wilhelmy plate method curve of Fig. 3.36(b) suggests that in the region
$(3 \rightarrow 7) \times 10^{-3}$ mol dm^{-3} this is correct but that some new factor has intruded

above this higher concentration. This factor is the formation of micelles, which are aggregates of about 100 long chain ions with a slightly smaller number of counter ions, and have a small residual charge. The actual size and degree of counter ion binding of the micelles will vary from tenside to tenside. The concentration at which this aggregation starts to occur is known as the critical micelle concentration (CMC), and is normally quoted as 8×10^{-3} mol dm^{-3} for SDS. It is interesting that the authors should have shown in Fig. 3.35 that the Wilhelmy plate method is the more reliable, and yet in Fig. 3.36(a) have marked the CMC as that derived from the drop volume method, which is inconsistent.

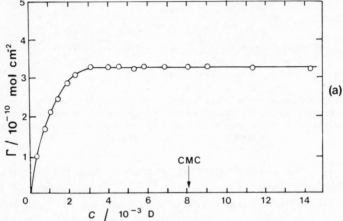

Fig. 3.36(a) – The adsorption isotherm for SDS at the air-solution interface at 298.15 K, from tracer studies.

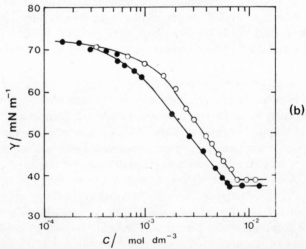

Fig. 3.36(b) – Surface tension vs concentration curves for SDS solutions at 298.15 K. –o– drop volume method, –•– Wilhelmy plate method. (From Tajima et al [52]).

On the basis of this micellar model, the apparent constancy of both Γ and γ at concentrations above the CMC, which at first sight would seem to defy the predictions of the Gibbs equation, is readily explained. In Fig. 3.36 the concentration scale referes to the total concentration of SDS, whereas in the Gibbs equation the concentration is that of the monomeric surface active ions, which is assumed constant above the CMC, thus γ and Γ remain constant also.

The micellar phenomenon is of course found with all classes of tenside with sufficiently large hydrophobic tails and hydrophilic head groups, and similar results [54] for various ethylene oxide non-ionic tensides are shown in Fig. 3.37. The subject, however, is really a colloidal phenomenon and cannot be considered further here.

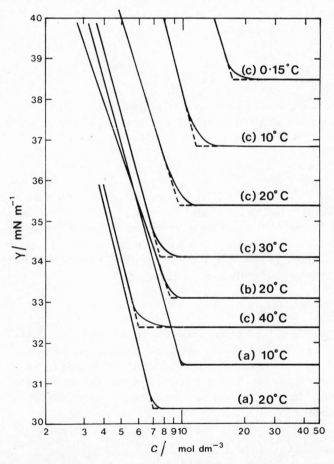

Fig. 3.37 – Surface tension *vs* concentration curves for non-ionic tensides in water.
(a) $C_{10}H_{21}O(CH_2CH_2O)_4H$, (b) $C_{10}H_{21}O(CH_2CH_2O)_5H$, (c) $C_{10}H_{21}O(CH_2CH_2O)_6H$.
(From Garrett [54]).

3.11.2 Non-electrolytes at the liquid-liquid interface

The Gibbs equation may also be used to predict the behaviour of interfacially active materials at the oil-water interface. An interesting example of this is the investigation of the effects of 1-octanol on the interfacial tension of the dodecane-water system [55]. Because of the many species present, the Gibbs equation no longer has a simple form. For two immiscible liquids, 1 and 2, and a solute, 3, distributed between them, it has the form

$$-\mathrm{d}\gamma = [\Gamma_3 - \Gamma_2 x_3^\beta/x_2^\beta - \Gamma_1 x_3^\alpha/x_1^\alpha]\ \mathrm{d}\mu_3 \qquad (3.126)$$

where α denotes the phase mainly of component 1, β that mainly of component 2, and the x values are mole fractions. For the system Aveyard and Briscoe [55] used, it is possible to put $x_3^\alpha = 0$, and since $x_3^\beta \sim 3 \times 10^{-3}$ in the most concentrated solution investigated, then (3.126) may be written

$$\frac{1}{\Gamma_3} = A = -\frac{kT}{x}\frac{\mathrm{d}x}{\mathrm{d}\gamma} \qquad (3.127)$$

where A is the area occupied by one molecule in the interfacial film. It has been shown that an empirical equation of state can be written which describes many such films, known as the Schofield-Rideal equation [56]

$$\Pi(A - A_0) = ikT \qquad (3.128)$$

where the surface pressure, $\Pi = \gamma_0 - \gamma$, that is, the difference in interfacial tensions between that for the simple 1,2 system, γ_0, and that when the component 3 is also present, γ. The term A_0 is the molecular 'co-area', of which more in a moment, and i is a constant. In the particular case when $i = 1$, equation (3.128) reduces to the form known as Volmer's equation, derived originally to describe two-dimensional gaseous films (see p. 121)

$$\Pi(A - A_0) = kT\ . \qquad (3.129)$$

In this equation A_0 is essentially analogous to the term b in the three-dimensional van der Waals equation of state for a real gas. Thus A_0 is the 'excluded area', that is to say the area of the interface which is unavailable to one molecule by the presence of another. From Fig. 3.38 it is obvious that for the pair of molecules represented, the centre of molecule B cannot get nearer than the distance $2r$ to the centre of molecule A, thus the excluded area should be

$$A_0 = \pi(2r)^2/2 = 2\pi r^2\ , \qquad (3.130)$$

in other words, twice the cross-sectional area of one molecule. This treatment may be termed the 'hard disk two-dimensional gas model'.

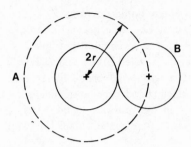

Fig. 3.38 — Definition of excluded area. The centre of molecule B cannot get nearer than the distance $2r$ to the centre of molecule A.

Using Aveyard and Briscoe's data gives apparently a very good fit to Volmer's equation as shown in Fig. 3.39 and the value of $A_0 = 0.24$ nm^2 over the experimental range investigated. They consider this value inconsistent with the above interpretation in terms of the hard-disk two-dimensional gas model because it is similar in magnitude to the alcohol A_0 values obtained from studies of insoluble monolayers, where it is interpreted as being the area occupied by one molecule. We will consider this point again later.

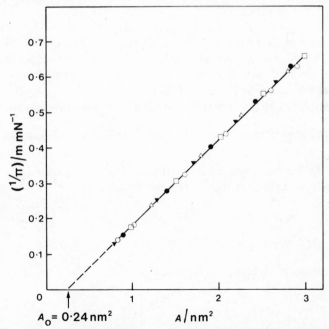

Fig. 3.39 — Application of Volmer's equation (3.120) to adsorbed films of l-octanol at the dodecane-water interface. Data calculated using the constants from Table 3 of Aveyard and Briscoe [55]. □ 308.15 K, △ 306.65 K, ○ 303.15 K, ● 300.65 K, ▼ 298.15 K, ■ 295.65 K.

3.11.3 The surface of binary liquid mixtures

Although this topic is of less wide general interest than those just discussed, but because in principle it is the simplest example to consider from a theoretical point of view, some space devoted to the subject is justified.

It is usual to start with what probably represents an over-simplification, namely that the interfacial layer may be treated as monolayer, in fact just as we did without stating it in the previous section, and that all the liquid that lies beneath this monolayer may be treated as bulk liquid mixture. If we let the surface area be A, then

$$A = \Sigma_i n_i^s \left[\frac{\partial A}{\partial n_i^s} \right]_{T,V^s,\gamma,n_j^s} \tag{3.131}$$

where n_i^s here defines the number of moles in a monolayer. The partial molar surface area term $\left[\dfrac{\partial A}{\partial n_i^s} \right]_{T,V^s,\gamma,n_j^s}$ may be approximated by the area occupied by one mole in a close-packed monolayer, a_i. Thus for a binary system

$$n_1^s a_1 + n_2^s a_2 = A \quad . \tag{3.132}$$

In order to derive an expression for the surface tension, suitable relations must be developed for the surface and bulk chemical potentials of a species, and these equated at equilibrium. The easiest case to consider is that of the perfect solution, which has been considered by Defay $et\ al$ [57].

In an ideal solution, the chemical potential μ_i^l of the component i is

$$\mu_i^l = \mu_i^{\ominus l} + RT \ln x_i^l \tag{3.133}$$

where x_i^l is the bulk solution mole fraction and $\mu_i^{\ominus l}$ the standard chemical potential. Whilst we have previously written for the surface chemical potential

$$\mu_i^s = \left[\frac{\partial A}{\partial n_i^s} \right]_{T,V^s,n_j^s,A} \tag{2.41}$$

using the properties of partial derivatives we may write

$$\mu_i^s = \left[\frac{\partial A^s}{\partial n_i^s} \right]_{T,V^s,n_j^s,\gamma} - \left[\frac{\partial A^s}{\partial A} \right]_{T,V^s,n_i^s} \left[\frac{\partial A}{\partial n_i^s} \right]_{T,V^s,n_j^s,\gamma} \tag{3.134}$$

which using equation (2.10) and the previous notation gives

$$\mu_i^s = \left[\frac{\partial A}{\partial n_i^s}\right]_{T,V^s,n_j^s,\gamma} - \gamma a_i \quad . \tag{3.135}$$

Now we may make the substitution

$$\left[\frac{\partial A^s}{\partial n_i^s}\right]_{T,V,n_j^s,\gamma} = \mu_i^{\ominus,s} + RT \ln x_i^s \tag{3.136}$$

and therefore

$$\mu_i^s = \mu_i^{\ominus,s} + RT \ln x_i^s - \gamma a_i \quad . \tag{3.137}$$

By equating chemical potentials at equilibrium we have

$$\mu_i^{\ominus,l} - \mu_i^{\ominus,s} = RT \ln \frac{x_i^s}{x_i^l} - \gamma a_i \quad . \tag{3.138}$$

If the solution is perfect then it must be ideal at all concentrations, therefore $\mu_i^{\ominus}$ must be that for a pure component, μ_i^0, and if we write $x_i^s = x_i^l = 1$ in equation (3.138) we have

$$\mu_i^{0,l} - \mu_i^{0,s} = \gamma_i^0 a_i \tag{3.139}$$

where γ_i^0 is the surface tension of the pure liquid, and if this is combined with equation (3.138) then

$$\gamma = \gamma_i^0 + \frac{RT}{a_i} \ln \frac{x_i^s}{x_i^l} \quad . \tag{3.140}$$

This equation works well for binary liquid mixtures which might be expected to behave ideally. These are usually molecules which are similar chemically in size: mixtures such as D_2O/H_2O, MeOH/EtOH and $\Phi Cl/\Phi Br$ for example. When molecules are of very different sizes then a more sophisticated approach is necessary.

For certain systems, Prigogine and Maréchal [58] have developed equations which relate the surface tension of mixtures of two components to the same parameters as in the simpler treatment, but also to the ratio of sizes of the molecular species. By a statistical thermodynamic approach to a system containing N_1 monomer molecules and N_2 chain-like r-mer molecules they obtain

$$\gamma = \gamma_1^0 + \frac{RT}{a_1} \left[\ln\left(\frac{\phi_1^s}{\phi_1^l}\right) + \frac{r-1}{r} (\phi_2^s - \phi_2^l) \right]$$

$$\text{(3.141)}$$

$$= \gamma_2^0 + \frac{RT}{ra_1} \left[\ln\left(\frac{\phi_2^s}{\phi_2^l}\right) + (r-1)(\phi_2^s - \phi_2^l) \right]$$

where a_1 is the molar surface area of the monomer, and the volume fractions are defined by

$$\phi_1 = \frac{N_1}{N_1 + rN_2} \qquad \phi_2 = \frac{rN_2}{N_1 + rN_2}$$

This approach has been used by Ono and Kondo [59] for systems of benzene and dibenzyl. The results of comparing theory and experiment are shown in Fig. 3.40, where $T = 60°C$, $r = 2$ and $a_1/L = 0.2894$ nm², and the agreement is obviously good. It should be noted that putting $r = 1$ in (3.141) gives the perfect solution, equation (3.140).

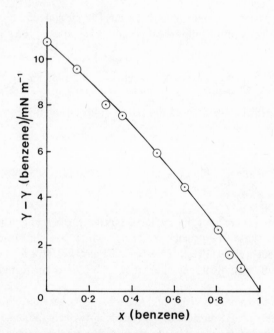

Fig. 3.40 – Comparison of the predictions of the Prigogine and Marechal equation (3.141) full line, with experimental results, –o–, for the benzene-dibenzyl system. (After Ono and Kondo [59]).

3.11.4 Insoluble surface films

The Benjamin Franklyn experiment on Clapham Common pond was an example of the study of an insoluble monolayer.[†] In this case the material in the surface film was not in thermodynamic equilibrium with the aqueous subphase. It is possible when the materials are sufficiently insoluble to approximate the film to a closed two-dimensional phase whose properties we may adjust by altering the area upon which it is spread. Many insoluble materials may be spread, even though they are solids, by using their solution in a volatile solvent. For example octadecanoic acid, which is a solid at room temperature, may be spread as a two-dimensional film by dissolving it in ultra-pure pentane, and spreading a small volume of solution on the ultra-clean water surface. The pentane evaporates from the spread solution leaving an insoluble monolayer of octadecanoic acid. It is perfectly possible to measure the surface pressure, Π, of a film spread on a known area, A, by the Wilhelmy plate technique, using the relationship $\Pi = \gamma_0 - \gamma$, where γ_0 is the surface tension of the clean surface and γ that of the film covered surface. However, wetting problems frequently cause contact angle complications in the use of the Wilhelmy technique.

The first workable alternative was due to Langmuir [60], and as modified by Adam and Jessop [61] forms the basis of most modern experiments of this type. Fig. 3.41 shows diagrammatically the principle of operation of such a trough.

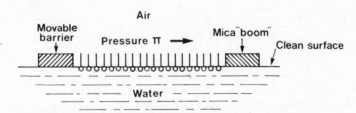

Fig. 3.41(a) – Diagrammatic representation of the principle of measurement of the Langmuir-Adam trough, in which the force due to surface pressure on the mica-boom is balanced by the force applied by a torsion wire suspension to which the boom is attached. The movable barrier is used to compress the film.

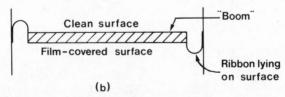

Fig. 3.41(b) – Adam's modification permitting the sealing of the boom to the edge of the trough whilst still enabling pressure measurements to be made.

[†] see the *Appendix* 3.12

The insoluble film is spread between the movable barrier and mica boom, both of which, together with the glass trough containing the aqueous sub-phase, are rendered hydrophobic by coating with either purified paraffin wax or PTFE. The area available to the insoluble monolayer may be altered by moving the movable barrier, and the surface pressure measured using the torsion wire balance attached to the mica boom. Adam used metal ribbon attached to the boom both to retain flexibility, and therefore not to interfere seriously with surface pressure measurement, as well as preventing the spread of the monolayer around the edges of the boom. Guastalla's further modification [62] of replacing the ribbons with silk threads covered with purified petroleum jelly is probably even more frequently used than ribbon these days. However, the experiments are still comparatively difficult to perform well, and detailed accounts can be found in the original references as well as in the books by Adam and Gaines quoted in the bibliography at the end of the chapter.

Since this book is primarily concerned with principles there is not the space to review all the experimental data on insoluble monolayers, so only a few specific examples will be given which illuminate the structure and characteristics of the monolayers. Of the fatty acids investigated in detail, tetradecanoic acid on 0.01 mol dm^{-3} hydrochloric acid illustrates well the variety of two-dimensional states that are conceivable.

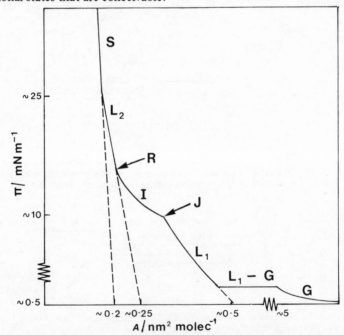

Fig.3.42 – Composite Π-A isotherm for tetradecanoic acid on 0.01 mol dm^{-3} HCl at ~25°C. Idealised diagram, with non-uniform scales, showing various surface states.

Fig. 3.42 is a composite force-area ($\Pi - A$) diagram and is not drawn with uniform scales. If the state marked G is the 2-d gaseous state and S the 2-d solid state, then exact analogies with conventional 3-d transitions become impossible for the intervening region. If $L_1 - G$ represents the transition between 2-d gas and 2-d liquid, the remaining regions must represent various forms of 2-d liquid. These are known as L_1, the liquid-expanded state, L_2, the liquid-condensed state and I the intermediate. It is possible to propose molecular models which would account for all of these. On the basis of proposals by Mittlemann and Palmer [63], Langmuir [48] and Adamson (see bibliography) it is possible to represent the transition from S to G as in Fig. 3.43.

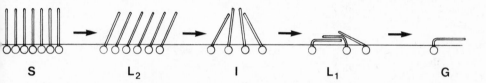

S L_2 I L_1 G

Fig. 3.43 – Model of 2-d film expansion: $S \rightarrow L_2 \rightarrow I \rightarrow L_1 \rightarrow G$.

The 2d-solid like film S is said to approximate to a close-packed arrangement of molecules. There is some discussion about what this means (the representation in the Fig. 3.43 is too simple) and we will return to the subject later, but for the present purposes it is sufficient to consider the molecules as being compressed to some limiting molecular area, A_0. In the case of tetradecanoic acid we can imagine the polar head groups being in the aqueous phase surrounded by water molecules, and the hydrocarbon tails out of the water, in contact with other hydrocarbon chains. It is the fact that the hydrocarbon tail interaction is greater than the hydrocarbon-water interaction that causes this particular orientation, and the insolubility of the film.

The transition from $S \rightarrow L_2$ involves an increase in A, possible inclination of the hydrocarbon tails to achieve maximum interaction, and an increased amount of water between the head groups, but still the film may be considered as continuous, as it was in the S state. The L_2 film will be very viscous and semi-solid perhaps. However, the $L_2 \rightarrow I$ transition was considered by Langmuir to represent the breakup of the continuous film into 'micelles' or clusters of molecules, which may be considered small enough to act like the kinetic units of an imperfect 2-d gas. The molecular area within a cluster will be less than the average value obtained experimentally, and the unit will be held together by the hydrocarbon chain interaction.

Further expansion of the film results in the L_1 state, with considerably increased freedom for the hydrocarbon tails following the breakdown of the 'micelles'. Mittlemann and Palmer [63] assumed a model in which there is a statistical distribution amongst the various possible configurations differing in the number, l', of the CH_2 groups of the hydrocarbon tail on the surface. The energy of a given configuration is

$$E = (\Pi + w) d \, l' + \epsilon' \tag{3.142}$$

where if d is the width of the chain, and w is a term related to the interaction energy of a CH_2 group with the water surface compared with its interaction with other CH_2 groups. The quantity ϵ' is the required work to rotate a C-C bond. Thus $\Pi d l'$ is the work the chain does in compressing the film and wdl' all the other energy changes except ϵ'. Using Boltzmann's equation they derive an expression

$$A = A_0 + \frac{d \sum\limits_{0}^{1} l' \exp\left(-(\Pi\, d\, l' + \epsilon')/kT\right)}{\sum\limits_{0}^{1} \exp\left(-(\Pi\, d\, l' + \epsilon')/kT\right)} \tag{3.143}$$

if the term wdl' may be neglected. With this model they obtained fair agreement for cis-9-octadecenoic acid (oleic acid) taking $A_0 = 0.16 \text{ nm}^2$, $d = 0.46$ nm and $l = 2.032$ nm, but these values are not without objection.

Table 3.7 shows that the model is not really satisfactory, being too simple a model of liquid structure and neglecting any change in head group interactions with separation.

Table 3.7
Comparison of experimental and calculated values of A for liquid cis-9-octadecenoic acid films (From Mittlemann & Palmer) [63].

$\Pi/\text{mN m}^{-1}$	2	4	6	8	10	12
A_{calc}/nm^2	0.57	0.51	0.46	0.41	0.38	0.34
A_{expt}/nm^2	0.52	0.485	0.46	0.44	0.42	0.405

Langmuir [48] approached the problem of the L_1 and I films in a very different way, considering that it should be possible to treat an insoluble monolayer as a duplex film with different upper and lower surfaces. The lower aqueous surface would be entirely —COOH groups and the upper essentially a normal paraffinic hydrocarbon surface. He noted that the shape of L_1 force-area curves as shown in a approximate way in Fig. 3.42, and the actual experiments [64] for the phase shown in Fig. 3.44, are almost exactly rectangular hyperbolas which could be represented by

$$(\Pi - \Pi_0)(A - A_0) = kT \qquad (3.144)$$

where $\Pi_0 = 11.2$ mN m^{-1}, and $A_0 = 0.12 + 0.00178\,(T - 273.15)$ nm^2.

The quantity Π_0 represents the spreading tension for the hydrocarbon part of the acid molecules, thus

$$\Pi_0 = \sigma^{OWV} = \gamma^{WV} - \gamma^{OW} - \gamma^{OV} \qquad (3.145)$$

For the region $20 > \Pi > 2$ mN m^{-1}, he obtained a remarkably good fit with the difference between calculated and experimental values having a standard deviation of 0.21 mN m^{-1}. However, the values used for Π_0 and A_0 present problems. For example Π_0 is independent of temperature, although fairly near the value of σ for hexadecane on water at 20°C, that is, -9.3 mN m^{-1}. On the other hand the values for A_0 are much less than the limiting areas at zero pressure determined experimentally, which are of the order of 0.20 nm^2 at 20°C, compared with the value of 0.156 nm^2 predicted for A_0 from Langmuir's data. Furthermore, as both Langmuir and Adam recognised, the duplex film model is somewhat difficult to maintain at the lower pressure end of the region, and it is correspondingly difficult to see why Π_0 should be invariant.

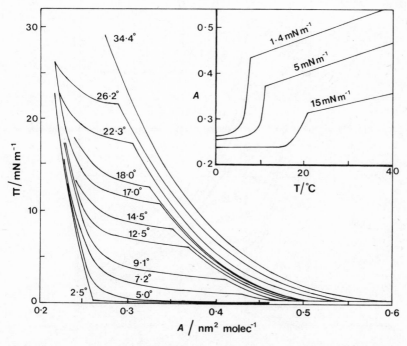

Fig. 3.44 – Π-A diagrams for tetradecanoic acid spread on 0.01 mol dm^{-3} HCl at various temperatures. (After Adam and Jessop [61]).

Langmuir tackled the problem of an I film by assuming that the film pressure in this region is the sum of that for the L_1 phase, taken as that at the point J in Fig. 3.42, and a contribution, Π_m, due to the surface micelles, each of which contain β monomer units. Thus

$$\Pi = \Pi_m + \Pi_J \quad . \tag{3.146}$$

If micelle formation disappears at pressures exceeding the point R in the figure and the micelles obey Volmers equation (3.129), then

$$\Pi_m = kT (A_J - A)/\beta A_J (A - A_R) \tag{3.147}$$

where A_J and A_R are the values of the molecular area, A, at the points J and R respectively. This equation is quite successful in predicting the shape of the force-area curve in the I region with $\beta = 13$, and in fact the success of equations (3.144) and (3.147) in reproducing the I and L_1 regions of the tetradecanoic acid diagram can be seen by comparing Langmuir's predictions shown in Fig. 3.45, with the experimental curves in Fig. 3.44.

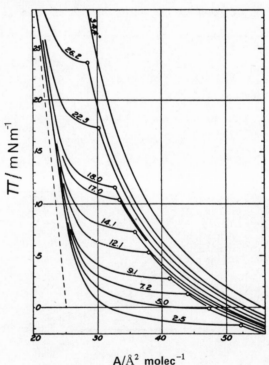

Fig. 3.45 – Langmuir's predicted tetradecanoic Π-A diagrams. (Reproduced from Langmuir [48]).

The surface micellar model also has the disadvantage that it requires that the micelles are all of one size, which would seem to be energetically impossible, but the alternative explanation suggested by Kirkwood [65] is that the I region is the result of the successive loss of rotational degrees of freedom as Π increases from the point J, until at R, the molecules no longer rotate. It is true to say that the real nature of I films is not really understood. Interfacial viscosities are low in the L_1 and I regions, and the interfacial potential increases continuously with increasing pressure, whereas the value $\mu_\perp$ remains almost constant, but since the latter is of the order of a tenth that associated with a –COOH group, interpretation is difficult and the role of water molecules in the structure is still unclear.

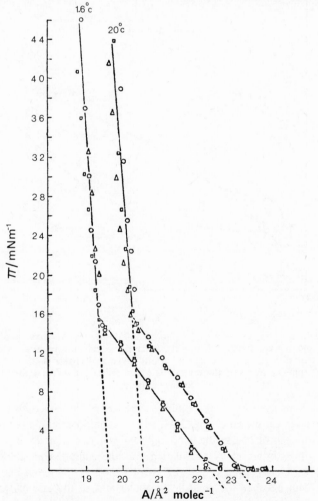

Fig. 3.46 – Π-A curves for octadecanoic acid on 0.01 mol dm^{-3} H$_2$SO$_4$ at 1.6°C and 20°C. (Reproduced with permission from Casilla *et al* [66]).

One attempt to understand the role of water in fatty-acid monolayers was that of Eley *et al* [66] who measured the force-area curves of octadecanoic acid monolayers as a function of temperature. Their results at 1.6°C and 20°C on an 0.01 mol dm^{-3} H$_2$SO$_4$ substrate[†] are shown in Fig. 3.46. These curves show only S and L$_2$ regions, and they noted that there was a difference between A$_0$, the area at zero pressure of the S phase line determined in their and many other measurements, and the values which would be predicted from 3-d crystal studies of solid octadecanoic acid, namely 0.182 to 0.184 nm^2 molec^{-1} (Fig. 3.47).

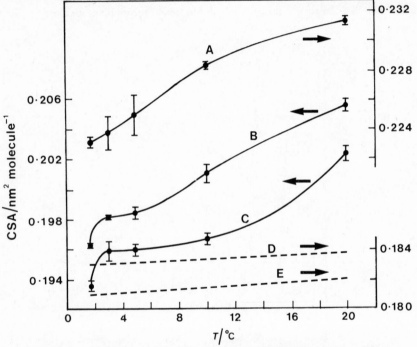

Fig. 3.47 — A$_0$ values for the solid (line B) and liquid (line A) octadecanoic acid films and the cross-sectional area (CSA) at the liquid/solid transition (line C) compared with the cross-sections of octadecanoic acid chains in the B (line D) and C (line E) forms of the crystal. The bars indicate standard errors. (From Casilla *et al* [66]).

The plots of the film compressibility measured at the S-L$_2$ transition of the S and L$_2$ phases are shown in Fig. 3.48, where a significant minimum is noticed near 4°C, which is of course the temperature of the maximum density of water. This the authors interpret by postulating that the compression of the L$_2$ film (envisaged as "close-packed heads" by Adam) involves the extension of water molecules from between the polar head groups, and, in the case of the S layer

[†]checked by private communication

(envisaged as "close-packed chains" by Adam), extrusion of the polar heads of alternate octadecanoic acid molecules into the interstitial volume of the water structure (Fig. 3.49). Since this volume must be at a minimum at 4°C, the minimum in the compressibility curve would make sense. However, the difference between the 3-d crystal cross-sections and the A_0 values of Fig. 3.46 indicate either a large free volume in the S phase or that some water is still present in the mono-molecular film, a view consistent with the fact that the S phase is still reasonably compressible.

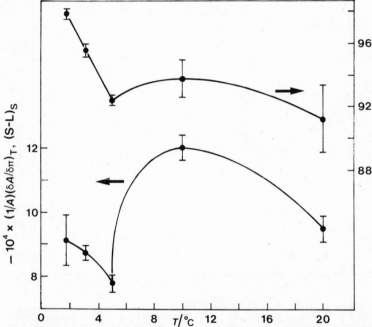

Fig. 3.48 – The compressibility at the solid/liquid intersection of the solid monolayer $(S\text{-}L)_S$ and the liquid monolayer $(S\text{-}L)_L$. The bars indicate the standard errors. (From Casilla *et al* [66]).

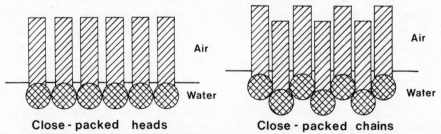

Close - packed heads   **Close - packed chains**

Fig. 3.49 – Diagrammatic representation of the "close-packed heads" and "close-packed chains" configurations of N. K. Adam. The position of the water-air interface as drawn has no real significance since it ignores the molecular characteristics of the aqueous phase.

The other interesting feature of their results is the unexplained decrease in compressibility above 10°C. Clearly their approach is worthy of extension to long-chain alcohols and amines, together with simultaneous measurement of other monolayer properties.

The extent of knowledge with other film materials is very similar, and results [67] for *cis*-9-octadecenol (oleyl alcohol) are shown in Fig. 3.50, which show similar comparisons between ΔV, $\mu_\perp$ and Π, as a function of area.

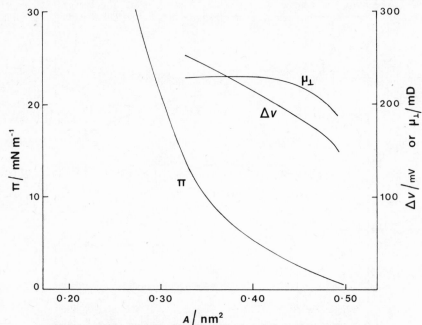

Fig. 3.50 – Surface pressure, Π, potential, ΔV, and perpendicular component of the surface dipole moment, $\mu_\perp$, for the expanded monolayer of *cis*-9-octadecenol (oleyl alcohol) on 0.001 mol dm^{-3} HCl at 21°C. (From Gaines [35], p. 168).

The amount of data in the literature on the viscosity of liquid and solid surface films is not large, and the patterns of behaviour found can be illustrated by the data of Boyd and Harkins [68]. At low surface pressures the viscosity of fatty acid monolayers is low and Newtonian, and increases with the chain length of the acid. However, at pressures above about 20 mN m^{-1} there is often a rapid increase in viscosity with the onset of non-Newtonian behaviour (Fig. 3.51).

The gaseous G state of insoluble monolayers is relatively easy to theorise upon, but much more difficult to experiment with. Obviously at very low surface pressures, and ideal 2-d gas law would be predicted:

$$\Pi A = kT \quad . \tag{3.148}$$

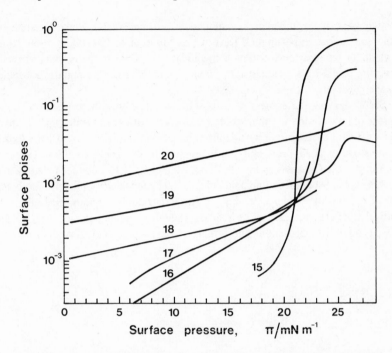

Fig. 3.51 – Interfacial viscosities (rotational) as a function of surface pressure for normal long chain acids on 0.01 mol dm^{-3} acid at 25°C. Numbers represent number of carbon atoms in the acid. (From Boyd and Harkins [68]).

If we imagine that intermolecular separations which would be equivalent to using $PV = nRT$ at 1 atm at 25°C are necessary for the 2-d state equation as well, the area would be 18.5 nm^2 molec^{-1} and the corresponding surface pressure 0.22 mN m^{-1}. Adam was able to show some degree of agreement with this equation at this order of area, but often larger areas are needed and experimentation is very difficult (see bibliography).

The obvious extension of equation (3.148) would be first to Volmer's equation

$$\Pi(A - A_0) = kT \tag{3.129}$$

but more significantly to the 2-d analogue of van der Waals equation for 3-d gas, usually known as the Hill-de Boer equation

$$(\Pi + \frac{\alpha}{A^2})(A - A_0) = kT \tag{3.149}$$

in which A_0 is the excluded area, analogous to van der Waals b, and α the analogue of the constant a. Experimental tests of this equation with insoluble monolayers are virtually non-existent, although the equation does have the obvious virtue of predicting the G-L_1 phase transition, which none of the other equations of state we have considered do.

Before leaving the subject of insoluble monolayers perhaps mention should be made of how steric and other molecular characteristics can produce pronounced affects on film behaviour. The introduction of *cis*-double bonds into the hydrocarbon chain or chain branching produces considerable film expansion. Compare for example the C_{18} acids in Fig. 3.52, *cis, cis, cis,* 9,12,15-octa-decatrienoic acid (linolenic), *cis,* 9-octadecenoic acid (oleic) [69], and 2-ethyl-hexadecanoic acid (2-ethyl palmitic) [70], with Fig. 3.46. Trans-double bonds produce comparatively little effect as can be seen in the comparison between the *cis*- and *trans*- forms of 13-docosenoic acid (erucic and brassidic) [71] shown in Fig. 3.53.

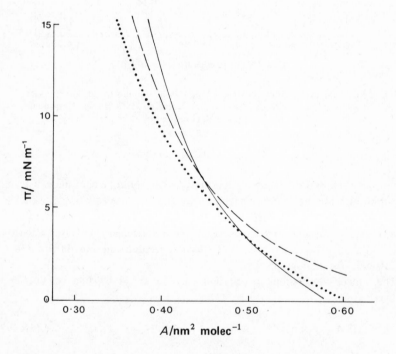

Fig. 3.52 – Π-A curves for various branched chain and unsaturated C_8 fatty acids at 15–20°C on 0.01 mol dm⁻³ HCl. (From Adam and Dyer [69], and Stenhagen [70]). – oleic acid, -- linoleic acid, ···· 2-ethyl palmitic acid.

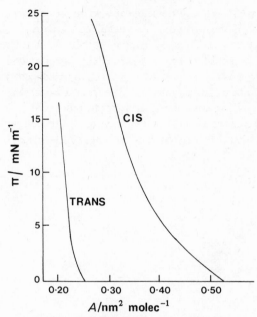

Fig. 3.53 – Π-A curves for the *cis-* and *trans-* forms of docosenoic acid. (From J. Marsden and E. K. Rideal [71]).

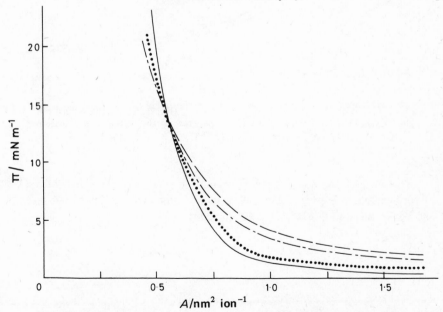

Fig. 3.54 – Π-A curves of $C_{18}H_{37}N(CH_3)_3^+$ ions on NaCl solutions of various concentration. ——2 mol dm^{-3}, · · · · 0.5 mol dm^{-3}, · — · — · 0.1 mol dm^{-3}, — — — 0.033 mol dm^{-3}, T = 21 ± 1°C. (From Davies [72]).

Remarkable expansion of the film can also be brought about by the presence of charge on the headgroup, thus octadecyltrimethyl-ammonium chloride monolayers may be considered as being essentially gaseous because of the long-range coulombic repulsion forces between the octadecyltrimethyl-ammonium ions in the monolayer [72, 73]. At high electrolyte concentration the film may be considered to be electrically neutral. The force/area curve is shown in Fig. 3.54 and the value of Π_0 in equation (3.144) is plotted in Fig. 3.55. It is interesting that the gaseous curve when $A > 1$ nm^2 agrees quite well with the predictions of equation (3.149) with an α value in the range $(74 \rightarrow 90) \times 10^{-20}$ J.

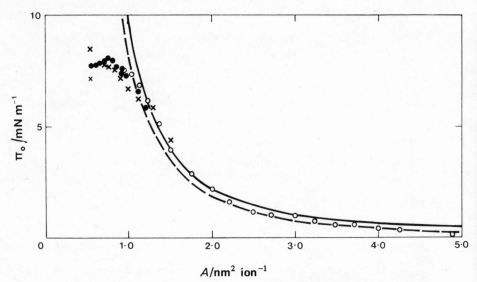

Fig. 3.55 – Cohesive pressure, $-\Pi_0$, of a film of $C_{18}H_{37}N(CH_3)_3^+$ ions as calculated by Davies [73], compared with curves predicted by the Hill-de Boer equation (3.149), $-\Pi_0 = \alpha/A^2$, with $\alpha = 74 \times 10^{-20}$ J, $--$, and 90×10^{-20} J, $\underline{\hspace{1cm}}$.

This chapter has not considered protein or other polymer films which are of considerable interest, and are well reviewed in the texts by Gaines, Adamson, and Davies and Rideal listed in the *Bibliography*.

3.12 APPENDIX
Benjamin Franklin
Phil. Trans. Roy. Soc. June 2nd, 1774 p. 144
Extract

Of the Stilling of Waves by Means of Oil. Extracted from Sundry Letters between Benjamin Franklin, L. L D. F. R. S. William Brownrigg, M. D. F. R- S. and the Rev. Mr. Farish.

Read at the Royal Society, June 2, 1774.

Extract of a Letter from Dr. Brownrigg to Dr. Franklin. dated Ormathwait, January 27, 1773.

BY the enclosed from an old friend, a worthy clergy-man at Carlisle, whose great learning and extensive knowledge in most sciences would have more distin-guished him, had he been placed in a more conspicuous point of view, you will find, that he had heard of your experiment on Derwent Lake, and has thrown together what he could collect on that subject; to which I have subjoined one experiment from the relation of another gentleman.

Extract of a Letter from the Rev. Mr. Farish, to Dr. Brownrigg.

I SOME time ago met with Mr. Dun, who surprised me with an account of an experiment you had tried upon the Derwent Water, in company with Sir John Pringle and Dr. Franklin. According to his representation, the water, which had been in great agitation before, was instantly calmed upon pouring in only a very small quantity of oil, and that to so great a distance round the boat as seemed incredible. I have since had the same accounts from others, but I suspect all of a little exag-geration. Pliny mentions this property of oil as known particularly to the divers, who made use of it in his days,

in

in order to have a more steady light at the bottom.* The sailors, I have been told, have observed something of the same kind in our days, that the water is always remarkably smoother, in the wake of a ship that has been newly tallowed, than it is in one that is foul.   Mr. Pennant also mentions an observation of the like nature made by the seal catchers in Scotland.   *Brit. Zool.* Vol. IV. *Article* Seal.   When these animals are devouring a very oily fish, which they always do under water, the waves alone are observed to be remarkably smooth, and by this mark the fishermen know where to look for them.   Old Pliny does not usually meet with all the credit I am inclined to think he deserves.   I shall be glad to have an authentic account of the Keswick experiment, and if it comes up to the representations that have been made of it, I shall not much hesitate to believe the old gentleman in another more wonderful phenomenon he relates of stilling a tempest only by throwing up a little vinegar into the air.

Extract of a Letter to Dr. Brownrigg from Dr. Franklin.

Dear Sir,   *London, Nov.* 7, 1773.

I THANK you for the remarks of your learned friend at Carlisle.   I had, when a youth, read and smiled at

---

Note by Dr. Brownrigg.

* Sir Gilfred Lawson, who served long in the army at Gibraltar, assures me, that the fishermen in that place are accustomed to pour a little oil on the sea, in order to still its motion, that they may be enabled to see the oysters laying at its bottom, which are there very large, and which they take up with a proper instrument. This Sir Gilfred had often seen there performed, and said the same was practised on other parts of the Spanish coast.

Pliny's account of a practice among the seamen of his time, to still the wave in a storm by pouring oil into the sea; which he mentions, as well as the use made of oil by the divers; but the stilling a tempest by throwing vinegar into the air had escaped me. I think with your friend, that it has been of late too much the mode to slight the learning of the ancients. The learned, too, are apt to slight too much the knowledge of the vulgar. The cooling by evaporation was long an instance of the latter. This art of smoothing the waves by oil is an instance of both.

Perhaps you may not dislike to have an account of all I have heard, and learnt, and done in this way. Take it if you please as follows.

In 1757, being at sea in a fleet of 96 sail bound against Louisbourg, I observed the waves of two of the ships to be remarkably smooth, while all the others were ruffled by the wind, which blew fresh. Being puzzled with the differing appearance, I at last pointed it out to our captain, and asked him the meaning of it. "The cooks," says he, "have, I suppose, been just emptying their greasy water through the scuppers, which has greased the sides of those ships a little;" and this answer he gave me with an air of some little contempt, as to a person ignorant of what every body else knew. In my own mind I at first slighted his solution, though I was not able to think of another, but recollecting what I had formerly read in Pliny, I resolved to make some experiment of the effect of oil on water, when I should have opportunity.

Afterwards being again at sea in 1762, I first observed the wonderful quietness of oil on agitated water, in the swinging glass lamp I made to hang up in the cabin, as described

described in my printed papers*. This I was continually looking at and considering, as an appearance to me inexplicable. An old sea captain, then
a passenger with me, thought little of it, supposing it an effect of the same kind with that of oil put
on water to smooth it, which he said was a practice of the Bermudians when they would strike fish,
which they could not see, if the surface of the water was
ruffled by the wind. This practice I had never before
heard of, and was obliged to him for the information;
though I thought him mistaken as to the sameness of
the experiment, the operations being different as well
as the effects. In one case, the water is smooth till the
oil is put on, and then becomes agitated. In the other
it is agitated before the oil is applied, and then becomes
smooth. The same gentleman told me, he had heard it
was a practice with the fishermen of Lisbon when about
to return into the river (if they saw before them too
great a surf upon the bar, which they apprehended
might fill their boats in passing) to empty a bottle or
two of oil into the sea, which would suppress the break-
ers, and allow them to pass safely. A confirmation of
this I have not since had an opportunity of obtaining;
but discoursing of it with another person, who had
often been in the Mediterranean, I was informed, that
the divers there, who, when under water in their busi-
-ness, need light, which the curling of the surface inter-
rupts by the refractions of so many little waves, let a
small quantity of oil now and then out of their mouths,
which rising to the surface smooths it, and permits the
light to come down to them. All these informations I

at times revolved in my mind, and wondered to find no mention of them in our books of experimental philosophy.

At length being at Clapham, where there is, on the common, a large pond, which I observed one day to be very rough with the wind, I fetched out a cruet of oil, and dropt a little of it on the water. I saw it spread itself with surprising swiftness upon the surface; but the effect of smoothing the waves was not produced; for I had applied it first on the leeward side of the pond, where the waves were largest, and the wind drove my oil back upon the shore. I then went to the windward side where they began to form; and there the oil, though not more than a teaspoonful, produced an instant calm over a space several yards square, which spread amazingly, and extended itself gradually till it reached the lee side, making all that quarter of the pond, perhaps half an acre, as smooth as a looking-glass.

After this I contrived to take with me, whenever I went into the country, a little oil in the upper hollow joint of my bamboo cane, with which I might repeat the experiments, as opportunity should offer, and I found it constantly to succeed.

In these experiments, one circumstance struck me with particular surprise. This was the sudden, wide, and forcible spreading of a drop of oil on the face of the water, which I do not know that any body has hitherto considered. If a drop of oil is put on a highly polished marble table, or on a looking-glass that lies horizontally, the drop remains in its place, spreading very little. But when put on water, it spreads instantly many feet round, becoming so thin as to produce the prismatic

tic

tic colours, for a considerable space, and beyond them
so much thinner as to be invisible, except in its effect
of smoothing the waves at a much greater distance.
It seems as if a mutual repulsion between its particles
took place as soon as it touched the water, and a repulsion so strong as to act on other bodies swimming on
the surface, as straw, leaves, chips, &c. forcing them to
recede every way from the drop, as from a centre, leaving a large clear space. The quantity of this force,
and the distance to which it will operate, I have not
yet ascertained; but I think it a curious enquiry, and
I wish to understand whence it arises.

In our journey to the north, when we had the plea-
sure of seeing you at Ormathwaite, we visited the cele-
brated Mr Smeaton, near Leeds. Being about to show
him the smoothing experiment on a little pond near his
house, an ingenious pupil of his, Mr. Jessop, then present, told us of an odd appearance on that pond, which
had lately occurred to him. He was about to clean a
little cup in which he kept oil, and he threw upon the
water some flies that had been drowned in the oil.
These flies presently began to move, and turned round
on the water very rapidly, as if they were vigorously
alive, though on examination he found they were not
so. I immediately concluded that the motion was occasioned by the power of the repulsion above mentioned, and that the oil issuing gradually from the
spungy body of the fly continued the motion. He
found some more flies drowned in oil, with which the
experiment was repeated before us. To show that it
was not any effect of life recovered by the flies, I imi-
tated it by little bits of oiled chips and paper cut in the
form of a comma, of the size of a common fly; when

the stream of repelling particles issuing from the point
made the comma turn round the contrary way. This
is not a chamber experiment; for it cannot be well repeated in a bowl or dish of water on a table. A considerable surface of water is necessary to give room for
the expansion of a small quantity of oil. In a dish of
water, if the smallest drop of oil be let fall in the mid-
dle, the whole surface is presently covered with a thin
greasy film proceeding from the drop; but as soon as
that film has reached the sides of the dish, no more
will issue from the drop, but it remains in the form of
oil, the sides of the dish putting a stop to its dissipation
by prohibiting the farther expansion of the film.

Our friend Sir John Pringle, being soon after in Scotland, learned there, that those employed in the herring-
fishery could at a distance see where the shoals of her-
rings were, by the smoothness of the water over them,
which might possibly be occasioned, he thought, by
some oiliness proceeding from their bodies.

A gentleman from Rhode-island told me, it had been
remarked, that the harbour of Newport was ever smooth
while any whaling vessels were in it; which probably
arose from hence, that the blubber which they sometimes bring loose in the hold, or the leakage of their
barrels, might afford some oil, to mix with that water,
which from time to time they pump out to keep their
vessel free, and that some oil might spread over the
surface of the water in the harbour, and prevent the
forming of any waves.

3.13 BIBLIOGRAPHY

N. K. Adam, (1941) *The Physics and Chemistry of Surfaces,* 3rd Edition, Oxford University Press, Oxford.

A. W. Adamson, (1976) *Physical Chemistry of Surfaces.* 3rd Edition, Wiley, New York.

R. Aveyard and D. A. Haydon, (1973) *An Introduction to the Principles of Surface Chemistry,* Cambridge University Press, Cambridge.

J. T. Davies and E. K. Rideal, (1963) *Interfacial Phenomena,* 2nd ed., Academic Press, New York.

G. L. Gaines, Jr., (1966) *Insoluble Monolayers at Liquid-Gas Interfaces,* Interscience, New York.

J. F. Padday, (1969) in *Surface and Colloid Science,* ed. E. Matijevic, Wiley-Interscience, New York, Vol. 1, p. 39.

R. Parsons, (1954) in *Modern Aspects of Electrochemistry,* ed. J. O'M. Bockris and B. E. Conway, Butterworths, London, p. 103.

M. J. Sparnaay, (1972) *The Electrical Double Layer,* Pergamon Press, Oxford.

Monolayers, Memorial Symposium to N. K. Adam, ed. E. D. Goddard, Amer. Chem. Soc., Adv. Chem. Ser. No. 144, (1975).

3.14 REFERENCES

[1] Lord Kelvin (W. Thomson), (1871) *Phil. Mag.,* **42**, 448; see also (1870) *Proc. Roy. Soc. (Edinburgh)* **7**, 63.

[2] P. S. de Laplace, (1805) *Mechanique Celeste, Suppl. Au X Livre,* Coureier, Paris.

[3] F. Bashforth and J. C. Adams, (1883) *An Attempt to Test the Theory of Capillary Action,* Cambridge University Press.

[4] Lord Rayleigh (J. W. Strutt), (1915) *Proc. Roy. Soc. (London),* **A92**, 184.

[5] Lord Rayleigh, (1892) *Phil. Mag.,* **34**, 309.

[6] J. F. Padday, (1969) in *Surface and Colloid Science,* Ed. E. Matijevic, Wiley-Interscience, New York, Vol. 1, p. 39.

[7] L. Wilhemly, (1863) *Ann. Physik,* **119**, 177.

[8] T. W. Richards and L. B. Coombs, (1915) *J. Amer. Chem. Soc.,* **37**, 1656.

[9] J. Jurin, (1718) *Phil. Trans.,* **29-30**, 739.

[10] Hagen and Desians, quoted in Ref. [4].

[11] S. Sugden, (1921) *J. Chem. Soc.,* **1483** (see also reference 6).

[12] J. M. Andreas, E. A. Hauser and W. B. Tucker, (1938) *J. Phys. Chem.,* **42**, 1001.

[13] P. F. Levin, E. Pitts and G. C. Terry, (1976) *J. Chem. Soc. Faraday Trans. I,* **72**, 1519.

[14] J. F. Padday, A. R. Pitt and R. M. Pashley, (1975) *J. Chem. Soc. Faraday Trans. I,* **71**, 1919.

[15] J. F. Padday, (1979) *J. Chem. Soc. Faraday Trans. I*, **75**, 2827.

[16] L. du Noüy, (1919) *J. Gen. Physiol.*, **1**, 521.

[17] Ref. 6., p. 129.

[18] W. D. Harkins and H. F. Jordan, (1930) *J. Amer. Chem. Soc.*, **52**, 1751.

[19] W. D. Harkins and F. E. Brown, (1919) *J. Amer. Chem. Soc.*, **41**, 503.

[20] J. L. Lando and H. T. Oakley, (1967) *J. Colloid Interface Sci.*, **25**, 526.

[21] S. Sugden, (1930) *The Parachor and Valency*, Routledge, London, Table 80, p. 219.

[22] Ref. 6, p. 147.

[23] F. M. Fowkes, (1965) *Chemistry and Physics of Interfaces*, Amer. Chem. Soc., p. 1.

[24] J. G. Kirkwood and F. P. Buff, (1949) *J. Chem. Phys.*, **17**, 338.

[25] I. Prigogine and L. Saraga, (1952) *J. Chim. Phys.*, **49**, 399.

[26] E. A. Guggenheim, (1959) *Thermodynamics*, 4th ed. North Holland, Amsterdam, p. 195.

[27] F. C. Goodrich, (1969) in *Surface and Colloid Science*, Ed. E. Matijevic, Wiley-Interscience, New York, Vol. 1, p. 1.

[28] R. Von Eötvös, (1886) *Ann. Physik.*, **27**, 448.

[29] J. D. van der Waals, (1894) *Z. Phys. Chem.*, **13**, 657.

[30] R. Parsons, (1954) *Modern Aspects of Electrochemistry*, Ed. J. O'M. Bockris and B. E. Conway, Butterworths, London, p. 103.

[31] J. Guyot, (1924) *Ann. de Pysique*, **2**, 506, and also for example
    A. N. Frumkin, (1925) *Z. Physik. Chem. (Leipzig)*, **116**, 485.
    J. H. Schulman and E. K. Rideal, (1931) *Proc. Roy. Soc. (London)*, **A130**, 259.
    N. K. Adam and J. B. Harding, (1932) *Proc. Roy. Soc. (London)*, **A138**, 411.
    W. D. Harkins and E. K. Fischer, (1933) *J. Chem. Phys.*, **1**, 852.

[32] For the origins of this method see
    A. Volta, (1801) *Ann. Chim. Phys.*, **40**, 225 cited in N. K. Adam, *Physics and Chemistry of Surfaces*, 3rd ed., Oxford University Press, (1941) p. 308.
    Lord Kelvin, (1898) *Phil. Mag.*, **46**, 91.
    W. A. Zisman, (1932) *Rev. Sci. Instr.*, **3**, 369.
    H. G. Yamins and W. A. Zisman, (1933) *J. Chem. Phys.*, **1**, 656.

[33] G. Gouy, (1910) *J. Physique*, **9**, 457; (1917) *Ann. Physik*, **7**, 129.
    D. L. Chapman, (1913) *Phil. Mag.*, **25**, 475,
    But see also particularly:
    J. Th. G. Overbeek (1952) in *Colloid Science*, Ed. H. R. Kruyt, Elsevier, Amsterdam, Vol. 1, Ch. IV.

[34] O. Stern, (1924) *Z. Elektrochem.*, **30**, 508.

[35] G. L. Gaines, (1966) *Insoluble Monolayers at Liquid-Gas Interfaces*, Interscience, New York p. 89.

[36] J. T. Davies and E. K. Rideal, (1963) *Interfacial Phenomena*, 2nd ed., Academic Press, New York p. 252.

[37] N. W. Tschoegl, (1958) *Aust. J. Phys.*, **11**, 154.

[38] D. W. Criddle and A. L. Meader, Jr., (1955) *J. Appl. Phys.*, **26**, 838.

[39] R. J. Myers and W. D. Harkins, (1937) *J. Chem. Phys.*, **5**, 601.

[40] W. D. Harkins and J. G. Kirkwood, (1938) *J. Chem. Phys.*, **6**, 53, 298.

[41] I. Langmuir and V. J. Schaefer, (1937) *J. Amer. Chem. Soc.*, **59**, 2400.

[42] N. W. Tschoegl, (1962) *Kolloid Zeits. u. Zeits. für Polym.*, **181**, 19.

[43] A. Pockels, (1891) *Nature, 43*, 437.

[44] see *Hydrodynamics* by Sir Horace Lamb, 6th ed., Cambridge University Press, (1953) p. 459.

[45] Ref. 36, p. 274.

[46] R. Dorrestein, (1951) *Proc. Acad. Sci. Amsterdam*, **B54**, 250, 350; W. Eisenmenger, (1959) *Acoustica*, **9**, 327.

[47] W. D. Harkins, (1952) *The Physical Chemistry of Surface Films*, Reinhold, New York.

[48] I. Langmuir, (1933) *J. Chem. Phys.*, **1**, 756.

[49] J. W. McBain and C. W. Humphreys, (1932) *J. Phys. Chem.*, **36**, 300.

[50] J. W. McBain, G. F. Mills and T. F. Ford, (1940) *Trans. Faraday Soc.*, **36**, 931.

[51] D. J. Salley, A. J. Weith, Jr., A. A. Argyle and J. K. Dixon, (1950) *Proc. Roy. Soc. (London)*, **A203**, 42.

[52] K. Tajima, M. Murumatsu and T. Sasaki, (1970) *Bull. Chem. Soc. Japan*, **43**, 1991; K. Tajima, (1970) *Bull. Chem. Soc., Japan*, **43**, 3063.

[53] K. Sekine, T. Seimiya and T. Sasaki, (1970) *Bull. Chem. Soc., Japan*, **43**, 629.

[54] H. E. Garrett, (1972) *Surface Active Chemicals*, Pergamon, Oxford, p. 31.

[55] R. Aveyard and B. J. Briscoe, (1970) *Trans. Faraday Soc.*, **66**, 2911.

[56] R. K. Schofield and E. K. Rideal, (1925) *Proc. Roy. Soc. (London)*, **A109**, 57.

[57] R. Defay, I. Prigogine, A. Bellemans and D. H. Everett, (1966) *Surface Tension and Adsorption*, Longmans, London, p. 166.

[58] I. Prigogine and J. Marechal, (1952) *J. Colloid Sci.*, 7, 122.

[59] S. Ono and S. Kondo, (1960) *Handbuch der Physik*, Ed Flügge, Springer, Berlin, Vol. 10, p. 134.

[60] I. Langmuir, (1917) *J. Amer. Chem. Soc.*, **39**, 1848.

[61] N. K. Adam and G. Jessop, (1926) *Proc. Roy. Soc. (London)*, **A110**, 423.

[62] J. Guastalla, (1929) *Compts. Rend.*, **189**, 241.

[63] R. Mittelmann and R. C. Palmer, (1942) *Trans. Faraday Soc.*, **38**, 506.

[64] N. K. Adam and G. Jessop, (1926) *Proc. Roy. Soc. (London)*, **A112**, 362.

[65] J. G. Kirkwood, (1943) in *Surface Chemistry*, AAAS Publication No. 21 p. 157.

[66] R. Casilla, W. D. Cooper and D. D. Eley, (1973) *J. Chem. Soc. Faraday Trans. I,* **69**, 257.

[67] Ref. 35, p. 168.

[68] E. Boyd and W. D. Harkins, (1939) *J. Amer. Chem. Soc.,* **61**, 1188.

[69] N. K. Adam and J. W. W. Dyer, (1924) *Proc. Roy. Soc. (London),* **A106**, 694.

[70] E. Stenhagen, (1940) *Trans. Faraday Soc.,* **36**, 597.

[71] J. Marsden and E. K. Rideal, (1938) *J. Chem. Soc.,* 1163.

[72] J. T. Davies, (1951) *Proc. Roy. Soc. (London),* **A208**, 224.

[73] J. T. Davies, (1956) *J. Colloid Sci.,* **11**, 377.

CHAPTER 4

The description of solid surfaces

4.1 INTRODUCTION

Compared to the liquid surfaces described in the previous chapter, solid surfaces are in many ways much more difficult to treat from a theoretical point of view and also more difficult to quantify in certain important respects. We are so familiar with the physical appearance of so many solid objects that there is a tendency to think of their form as being a stable thermodynamic equilibrium, which it normally is not. If most solid geometries were the minimum energy configuration, then such processses as annealing and sintering would not be possible, and we know well the practical usefulness of both phenomena. For example a 12-inch steel ruler appears a stable object, but the very act of engraving the calibration markings on the surface of the ruler has increased the total surface energy of the system by increasing the surface area. Although we are not worried about the markings fading in the attempt to minimise the free energy of the system by decreasing the area, if the ruler were heated to a temperature just below its melting point, then it would be possible to see such a process occurring. The form of the ruler we are used to is a thermodynamically unstable state, yet no observable change occurs on the timescale we can conceive, for entirely kinetic reasons.

One of the other major problems in describing the solid state is that the surface tension of the solid is not susceptible to measurement. Values can be calculated, but direct measurement is not normally possible. This is a severe hindrance in characterising many solid surfaces, and methods of study have to take different routes. This chapter is concerned with how a solid surface may be adequately described from a theoretical point of view, some of the methods that can be used to investigate it, and some of the factors which modify its behaviour.

4.2 THE SURFACE ENERGY OF A SOLID
4.2.1 Surface tension and surface free energy
The newly formed surface of a liquid rapidly takes up an equilibrium conformation, whereas, as we have previously remarked, the same is not true of a solid surface. The latter is likely to have a considerable range of values of surface free

energy, varying from region to region on the surface, and also at any one point, unlike the surface of a liquid, the surface tension need not be the same in all directions. Let us suppose that the surface tension can be resolved into two directions at right angles, and that we can represent these partial surface tensions by γ^1 and γ^2. For an anisotropic solid, if the area is increased in two directions by dA_1 and dA_2, as shown in Fig. 4.1, then the total increase in available energy is given [1] by the reversible work done against the stress γ^1 and γ^2. Thus

$$d(AA^s)_{T,V,n} = \gamma^1 \, dA_1 + \gamma^2 \, dA_2 \tag{4.1}$$

where A^s denotes the available energy per unit area. If $\gamma^1 = \gamma^2 = \gamma$ then

$$\gamma = \frac{d(AA^s)_{T,V,n}}{dA} = A^s + A\left[\frac{dA^s}{dA}\right]_{T,V,n} \tag{4.2}$$

which can also be derived for an isotropic solid from equation (2.32).[†]

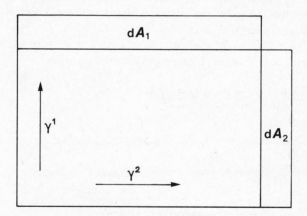

Fig. 4.1 – Resolution of the surface tensions of an anisotropic solid in two directions.

[†] The available energy for an open system may be written

$$dA = -SdT - PdV + \gamma dA + \Sigma_i \mu_i dn_i \tag{2.32}$$

and at constant T, V and n this yields

$$\gamma = \left(\frac{\partial A}{\partial A}\right)_{T,V,n} \tag{4.3}$$

Then $$dA = d(A^s A) = A^s dA + A dA^s \tag{4.4}$$

which on substitution in equation (4.3) gives equation (4.2).

In the case of a solid, however, $dA^s/dA \neq 0$ as a general rule, in fact it could only be so for an homotattic (ideal crystal-like) surface [2,3]. Thus only if the surface achieved some uniform equilibrium state would $A^s = \gamma$, and in all other real cases A^s and γ will be different from their equilibrium values and different from each other [4]. It cannot be legitimate therefore to call γ the specific surface free energy [5,6].

This problem can also be seen to have further implications. If we integrate equation (2.32) under conditions of constant T and V then

$$A = \gamma A + \Sigma_i \mu_i n_i \tag{4.5}$$

or

$$A^s = \gamma + \Sigma_i \mu_i \Gamma_i \quad . \tag{4.6}$$

Thus even if the conditions of an ideal surface could be maintained in making $dA^s/dA = 0$, then $\Sigma_i \mu_i \Gamma_i = 0$ must also be satisfied. The reader is recommended to consider this last point again after reading the section dealing with the calculation of surface energy.

This type of consideration of the surface in relation to Gibbs free energy raises certain problems [7]. If one starts with the equation

$$dG = -SdT + VdP + \gamma dA + \Sigma_i \mu_i dn_i \tag{2.33}$$

then

$$\left[\frac{\partial G}{\partial A} \right]_{T,P,n_i} = \gamma \tag{4.7}$$

and therefore for an isotropic solid

$$\gamma = \frac{\partial (G^s A)}{\partial A} = G^s + A \left[\frac{\partial G^s}{\partial A} \right]_{T,P,n_i} \tag{4.8}$$

which is analogous to equation (4.2). However, these concepts were introduced in this section using the ideas of reversible work, and the abnormal behaviour of G in this connection was illustrated in equations (2.25) and (2.38). If, on the other hand, we start with

$$dG = -SdT + VdP - Ad\gamma + \Sigma_i \mu_i dn_i \tag{2.40}$$

then

$$\left[\frac{\partial G}{\partial \gamma}\right]_{T,P,n_i} = -A \tag{4.9}$$

which if we expand analogously gives

$$\left[\frac{\partial (G^s A)}{\partial A}\right]_{T,P,n_i} = -A \frac{\partial \gamma}{\partial A}$$

or

$$G^s = -A \left[\frac{\partial \gamma}{\partial A} + \frac{\partial G^s}{\partial A}\right]_{T,P,n_i} \tag{4.10}$$

which is not similar in form to equation (4.2). On the other hand integrating equations (2.33) and (2.40) at constant T and P gives

$$G = \gamma A + \Sigma_i \mu_i n_i \tag{4.11}$$

$$G = -\gamma A + \Sigma_i \mu_i n_i$$

and therefore

$$G^s = \gamma + \Sigma_i \mu_i \Gamma_i$$
$$G^s = -\gamma + \Sigma_i \mu_i \Gamma_i \tag{4.12}$$

The concept of surface tension applied to a solid has important implications in relation to the Laplace equation (3.10). Let us consider for a moment a nearly spherical crystal such that we may write

$$\Delta P = 2\gamma/r \tag{3.10}$$

The pressure differential ΔP will result in compression of the crystal. Thus if β is the compressibility, then we may write the approximate relation

$$\frac{\Delta V}{V} = 3\frac{\Delta r}{r} = -\Delta P \beta \tag{4.13}$$

or

$$\Delta r = -2\beta\gamma/3 \quad . \tag{4.14}$$

This effect has been investigated [8] for magnesium (II) oxide, using X-ray powder diffraction measurements. Values of Δr near the expected value of

0.06 nm, corresponding to an 0.1% change in lattice distance, were found for a
60 nm radius crystallite using a calculated value for γ of 6.573 N m^{-1}. Rein-
vestigation of the situation by Guilliatt and Brett [9] indicates that any adsorbed
material such as water vapour produces lattice dilation, whereas in the absence
of an adsorbed film the expected lattice contraction is observed, the relative
magnitude of which increases with decreasing crystallite size, and is substantially
in agreement with the theoretical predictions of Anderson and Scholtz [10].
It is worth noting that surface dilation would imply a negative value for γ in
equation (4.14).

Shuttleworth [1] set out to show that equating of surface tension and
surface stress had very real conceptual problems for crystalline solids.[†] He
defined surface stress analogously to stress in bulk elasticity. If one imagines a
cut is made perpendicular to the surface of the crystal and extending only a
little way into it, then in order for the surface to achieve an equilibrium configuration, and no additional stresses to appear in the bulk of the crystal, then a
surface stress must be present, and could be achieved by local relaxation of the
crystal structure. Since this surface stress will have at least the symmetry of the
crystal face, then the surface tension can be taken to specify it when the surface
has a three-fold (or greater) axis of rotational symmetry. In this case the normal
stress components across all lines in the face are equal and all shear stresses are
zero, so that the normal stress components are equal to the surface tension.

In a face-centred cubic crystal the (111), (100) and (110) faces have six-fold,
four fold and two-fold axis of symmetry respectively. The shear component of
the surface stress will be zero for the (111) and (100) faces, but not for the (110)
face. Shuttleworth attempted calculations for (100) faces of inert gas crystals
and alkali halide crystals at 0K, expressing the surface energy as

$$U = U' + U''\tag{4.15}$$

where U' is the value derived from the surface energy of the crystal before any
rearrangement of the surface atoms relative to ideal positions occurs, and U''
that due to the surface energy decrease on relaxation. He concluded that the
surface stress was -130 mN m^{-1}. On the other hand Gibbs has shown that
the surface tension must be positive as a condition for stability, and Dunning
[11] quotes a value when reviewing the work of Shuttleworth [1] and Lennard-
Jones and Dent [12], of 155 mN m^{-1}. We shall return to this point when we
discuss the relaxation of a surface in section 4.2.2.

The mechanism by which surface stress could be relieved has been discussed
by Herring [13] and shown diagrammatically by Dunning [11].Suppose a cube of
ideal crystal is subjected to a compressive surface stress, then its deformation

[†]Gibbs has pointed out that the surface tension, or precisely the surface energy, is the work
necessary to form the surface, whereas the surface stress represents the work necessary to
stretch the surface.

(Fig. 4.2(a)) is equivalent to applying traction to each edge of the cube. This stress could be relieved by the presence of rows of dislocation just below the surface (Fig. 4.2(b)), since dislocations would have the effect of stretching a surface without increasing its area, or by vacancies actually in the surface (Fig. 4.2(c)). Such processes must involve increases in surface energy and the generation of surface heterogeneity.

(a)   (b)   (c)

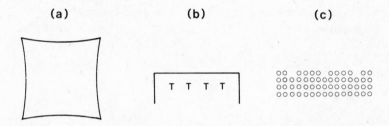

Fig. 4.2 — Herring's mechanism for the relief of surface stress by dislocations.
(After Dunning [11]).

4.2.2 Calculated surface energy values

Because of the impossibility of measuring surface tension values for solid surfaces, except in a few abnormal circumstances (see section 4.3) calculated surface energy values for solids take on a significance that they lack in the case of liquid interfaces. We have already referred to the early work in this field of particularly Lennard-Jones and Dent [12], but they, and the majority of other workers before about 1960, were considerably limited in terms of what they could attempt, by severe computational difficulties. The advent of the modern computer changed this situation, and we can illustrate the impact that these made by looking first at the (100) face of sodium chloride crystals.

The Lennard-Jones and Dent model, although extremely elegantly treated, was restricted by the fact that they considered an ideal crystal, in which the ionic positions at the surface were identical to those achieved in the bulk crystal. This is obviously extremely improbable, and the first attempt to understand relaxation processes was due to Verwey [14]. He assumed a bulk lattice spacing of 0.281 nm, and concluded that in the outermost layer the chloride ions moved out from the predicted ideal plane position to 0.286 nm from the plane of the ideal bulk crystal immediately beneath, and that the sodium ions moved inwards to 0.266 nm from the same plane, as illustrated in Fig. 4.3. Thus the plane of the outer chlorine ions was 0.020 nm farther out than the plane of the outer sodium ions, giving rise to a surface double layer. This differential effect is the result of the fact that the larger negative ions are more polarisable than the smaller positive ions, and that consequently there will be a larger induced electric dipole moment in the chloride ion than in the sodium ion. Thus the

outer chloride ions will move so as to increase the distance of the positive end
of the dipole from the plane of the sodium ions and decrease the distance of
separation of the negative end.

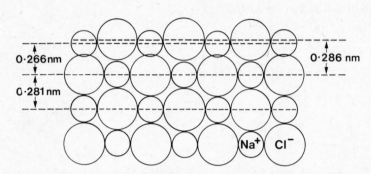

Fig. 4.3 – Verwey's single layer relaxation model of sodium chloride. In his
calculations, compared with a bulk crystal layer separation of 0.281 nm, the
outer sodium ions moved inwards by 0.015 nm and the outer chloride ions
outwards by 0.005 nm. The circles represent approximate ionic radii.

There would seem no logical reason for supposing that only the outer plane
should relax, and that some larger number of layers will in fact show significant
displacements. Benson *et al* [15], as part of a general study of relaxation in the
surfaces of alkali halide crystals [16], showed the outer five layers were all
significantly displaced from the ideal crystal position, and that only from the
sixth layer in from the surface onwards was it fair to consider the crystal as
bulk ideal crystal.

The method of calculation was to use numerical methods to minimise the
total energy of the system by allowing ions in the outer five layers to move to
new equilibrium positions. The interaction energy of two ions, i and j, was taken
to be

$$u_{ij} = u_{ij}^0 - e_i(\mathbf{r}_{ij}.\mu_j)r_{ij}^{-3} + e_j(\mathbf{r}_{ij}.\mu_i)r_{ij}^{-3} - 3(\mathbf{r}_{ij}.\mu_i)(\mathbf{r}_{ij}.\mu_j)r_{ij}^{-5} + (\mu_i.\mu_j)r_{ij}^{-7}$$

$$(4.16)$$

where e and μ denote charge and dipole moment respectively and $\mathbf{r}_{ij}$ is the
position vector of the ion i relative to the ion j, and

$$u_{ij}^0 = e_ie_jr_{ij}^{-1} - c_{ij}r_{ij}^{-6} - d_{ij}r_{ij}^{-8} + b_{ij}\exp(-r_{ij}/\rho) \qquad (4.17)$$

is the Born-Meyer form of the potential energy function and represents the
sum of coulombic, van der Waals and repulsive contributions. The other terms

in (4.16) are the energies of the charge-dipole and dipole-dipole interactions arising out of the polarisation of the ions, and the exponential term is the repulsive energy involving the 'hardness' parameter, ρ. Since the area occupied by an ion pair in the surface is $2a^2$, where a is the nearest neighbour separation in the bulk crystal, then the correction to the surface energy (compared with that of the ideal crystal) will be given by

$$\Delta\sigma_{(100)} = \Delta U_{min} / 2a^2 \tag{4.18}$$

where ΔU_{min} is the reduction in the energy of the crystal brought about by relaxation. Thus the true surface energy, $\sigma_{(100)}$, will be related to that of an ideal crystal, $\sigma^0_{(100)}$, by

$$\sigma_{(100)} = \sigma^0_{(100)} + \Delta\sigma_{(100)} \tag{4.19}$$

$$= 210.9 - 107.4 = 103.5 \text{ mJ m}^2$$

assuming the outer five layers relax. The significance of increasing the number of layers allowed to relax in the calculation can be seen in Fig. 4.4.

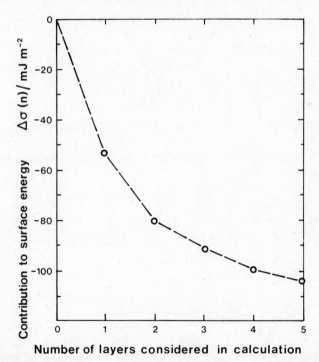

Fig. 4.4 – Values of the reduction in surface energy, $\Delta\sigma_{(100)}$, for the (100) face of NaCl, as a function of the number of layers assumed to relax. (From Benson *et al* [15]).

Benson *et al* [16] data on the positional changes of the ions was represented by them as shown in Fig. 4.5 for the case where the outer five layers are allowed to relax. The most significant factor is that whilst the chloride ions are increasingly displaced away from the surface as the surface is approached from the bulk of the crystal, the displacement of the smaller sodium ions alternates. However, the degree of disorder or distortion this relaxation process actually produces in the surface can be more readily seen in Fig. 4.6, in which the circles represent approximate values for the bulk crystalline ionic radii. It can be readily appreciated from the size of $\Delta\sigma_{100}$ and the positional shifts of Fig. 4.6, that crystal surfaces must generally be very different from bulk crystal planes, and this would be expected to produce considerable changes in adsorption behaviour from that predicted on the basis of ideal crystal geometry (see section 4.6.4).

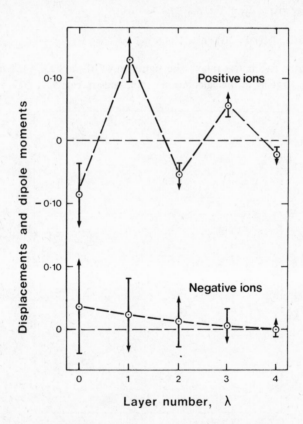

Fig. 4.5 – Equilibrium configuration of the first five layers, for the (100) face of NaCl. Displacements in units of *a*. Positive values indicate normal displacements outwards. The direction and magnitude of the induced dipole moments (Debye units) are indicated by the arrows, but for the negative ions they are scaled down by a factor of ten. (From Benson *et al* [15]).

Vacuum

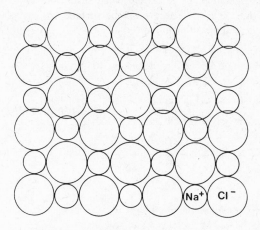

Bulk Crystal

Fig. 4.6 — Surface relaxation of the (100) face of sodium chloride crystals. The circles represent approximate ionic radii. (From Benson *et al* [15]).

Theoretical studies of relaxation at most crystal surfaces other than the alkali halides are still lacking. There is no real reason why any completely ionic crystals could not be treated in a similar way to that which Benson *et al* have adopted, but for crystals exhibiting a degree of covalent character, silver iodide for example, a different approach is needed. The most probable route to solving such problems would seem to lie through the use of wave-mechanical methods, but it is not clear at the moment how this could be done, and the authors are not aware of any successful attempt. Crystals of inert gases held together by van der Waals forces alone, can of course be dealt with in an analogous manner to that of Benson *et al*.

4.3 IMPERFECTIONS AND HETEROGENEITY

Just as it is known that an ideal bulk crystal is a practically unattainable state, the possibility of attaining an ideal surface, relaxed as described in the previous section, is even farther removed from reality. For a start such a crystal could not grow, and impurity levels have always been found to be finite, even if very, very low. The purpose of this section is to survey the types of defect that might be found, and to illustrate how they might influence the structure and behaviour of the crystal. As far as possible we will chose our examples from ionic solid surfaces, since they can be used to illustrate most types of behaviour, and only other types of solid surface when the scarcity of data makes this necessary.

4.3.1 Lattice defects

The surface of a crystalline ionic lattice can exhibit the same sorts of defect structure as the bulk crystal. There must be surface analogues of Frenkel and Schottky defects as illustrated in Fig. 4.7.

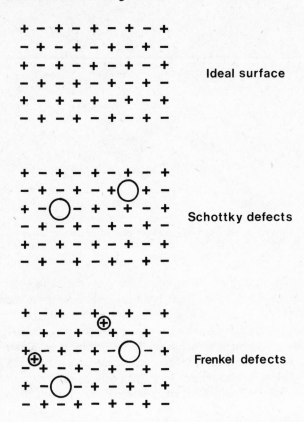

Fig. 4.7 — The surface of a univalent ionic crystal in the ideal state, and showing Schottky and Frenkel defects; plan view.

Schottky defects may be considered as being derived from the ideal crystal surface with vacancies in lattice positions, whereas Frenkel defects are the result of the movement of some of the ions, usually the cations because they are smaller, to interstitial positions leaving behind lattice vacancies. In the bulk crystal, if anions and cations are of the same size, Schottky defects predominate, whereas if the cations are much smaller than the anions then Frenkel defects are found, because it is now possible to accommodate the smaller ion in interstitial positions.

At the surface we effectively have a whole row of lattice vacancies necessarily present above the surface, therefore the creation of Schottky defects should be relatively easy, since there would be a vacancy on the surface waiting to receive an ion displaced from the surface layer. Subsequent relaxation would probably blur the distinction between 'in' and 'on' the surface. Likewise surface Frenkel defects may be easier to accommodate, because of surface relaxation, and the relaxation phenomenon itself will tend to blur the distinction between the two types of defect. For example, are the outer layer of sodium ions in Fig. 4.6 in interstitial positions, or is it only deviations from Fig. 4.6 that are defect structures? This secondary relaxation process around surface defects is virtually unstudied in the case of ionic crystals, but there are some interesting results for rare gas crystals.

Burton and Jura [17] adopted a very simple model for the representation of the (100) argon crystal surface. They assumed that all quantum effects could be neglected and that only potential energy need be considered. The total interaction potential was considered to be pair-wise additive and described by the Lennard-Jones 6-12 potential equation

$$U = \frac{A}{d^{12}} - \frac{B}{d^{6}} \tag{4.20}$$

where d is the distance between the pair. The surface relaxation of an argon crystal had been previously studied by Jura *et al* [18, 19], and they used as the basic structure the relaxed structure they had already determined. Relative to this the maximum additional displacements produced by adding an argon atom to the surface were all less than 2% of the interatomic spacing, and produced a change in the interaction energy of the added atom from 5.669 kJ mol^{-1} to 5.720 kJ mol^{-1}. They represented the relative movements of the argon atoms as shown in Fig. 4.8. The pattern of behaviour was in fact somewhat similar to that observed in calculations on relaxation in bulk argon crystals [20].

Fig. 4.8 (a) − Plan view of the relaxation of the surface layer of an argon crystal with an additional argon atom on top. Arrows indicate the directions of the displacements from the normal positions, and the added atom is represented by a square.
(b) − Side view of the surface relaxation. (From Burton and Jura [17]).

They also investigated the effect of creating a vacancy or Schottky defect in the surface, and found displacements of similar magnitude (Fig. 4.9). They calculated the Gibbs free energy of formation of a mole of vacancies to be

$$\Delta G = 5171 - 14.99 \, T \quad \text{J mol}^{-1}$$

when secondary relaxation around the vacancy is considered, and

$$\Delta G = 5322 - 14.99 \, T \quad \text{J mol}^{-1}$$

when it is not. This implies that at the melting point of argon there is one vacancy for every three hundred sites on the ideal flat (100) surface, or 2×10^{16} per m^2 of surface. Thus they concluded that although the apparent distortion around a defect of either type is small, it nevertheless appreciably alters the associated energies, and that consequently it is not neglectable in the calculation of surface properties.

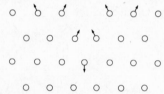

Fig. 4.9 – Side view of a section through the surface of an argon crystal, showing the relaxation of the atoms near a surface vacancy. Arrows indicate the directions of the displacements. (From Burton and Jura [17]).

4.3.2 Non-stoichiometry

Non-stoichiometry is a very common phenomenon particularly amongst metallic oxides. In fact it is present in a number of useful materials and sometimes confers properties which are a help, in the case of catalysts, or a hindrance, in the case of pigments. Titanium (IV) oxide is frequently found in forms that are oxygen deficient, and both rutile and anatase forms have been found where the formula may be written $TiO_{1.9 \to 2.0}$. This implies that there will be Schottky defects in the bulk crystal and a high probability of there being some at the surface, or sufficiently near it to locally alter the surface energy. To revert to the sodium chloride example of section 4.2.2 for a moment, it is obvious from Fig. 4.6 that the appearance of a Schottky defect in the surface layers of that structure will significantly alter the surface energy by producing displacement of the ions around the defect. There is no reason to suppose that other ionic materials will differ qualitatively in this respect, although on a quantitative basis they will be different. Thus most TiO_2 samples in say the rutile form are likely to show the results of Schottky defect in their surface behaviour.

The lower titanium (II) oxide [21] shows the effects of non-stoichiometry to a remarkable degree. This compound exhibits simple cubic structure of the extraordinary range of composition $TiO_{0.69 \to 1.33}$, and is stable above 900°C throughout the range. It has an extremely high concentration of Schottky defects, and at compositions either side of TiO is semiconductive. The material TiO itself conducts by an ionic mechanism but has 15% of both cation and anion sites vacant. $TiO_{1.33}$ on the other hand has 98% occupancy of the anion sites, but only 74% of the cation sites; consequently it behaves as a p-type semiconductor. At the other end of the composition range, $TiO_{0.69}$, the metal ion sites are 96% occupied whilst the oxide ion sites are only 66% occupied, and the material is an n-type semiconductor.

In fact there are usually considered [22] to be four types of non-stoichiometric lattices:

Type I,   metal present in excess due to anion vacancies; electrical neutrality being maintained by electrons trapped in the vacancies.

Type II,   metal excess due to interstitial metal ions; electrical neutrality being maintained by electrons trapped in the vicinity of the metal ions. This is obviously somewhat similar to Frenkel defects, but the metal ion vacancies are absent.

Type III, metal deficiency due to cation vacancies; electrical neutrality being maintained by oxidising some of the metal ions to a higher oxidation state.

Type IV,   metal deficiency due to interstitial anions; electrical neutrality again being maintained by oxidising some of the metal ions to a higher oxidation state.

All these types of behaviour are illustrated in Fig. 4.10.

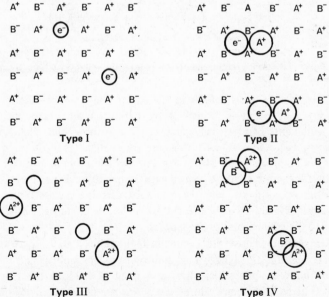

Fig. 4.10 – Types of defect in non-stoichiometric lattices.

The presence of any of these types of defect in the layers near the surface would produce local changes in surface energy; their presence in the outermost layer deserves further consideration. Whilst defects of types I and III with considerable local relaxation can be readily imagined, the possibility of finding types II and IV is much less probable. In the last two cases it would be expected that the relaxation process would result in the expulsion of the interstitial ion to a position above the outer plane, as represented in Fig. 4.11. Subsequent relaxation of the positions shown in the figure must occur, and has not been represented. Nevertheless the situation would have lower energy than trying to accommodate the additional ion in the surface layer.

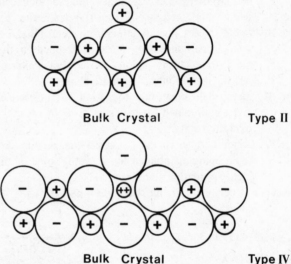

Fig. 4.11 – Possible rearrangements to reduce the energy of type II and IV surface defects. (The interstitial electron in type II is not shown).

The presence of such defects in the surface has implications in the field of heterogeneous catalysis, which is by definition a surface phenomenon. Bulk measurements of electrical conductivity also shows that they cause semi-conductivity, as noted earlier. In this case conductivity is in the range 10^{-5} to 10^{-9} Ω^{-1} cm compared with 10^3 to 10^5 Ω^{-1} cm for a metal, and is marked by the fact that the conductivity *increases* with temperature. Types I and II give rise to n-type semi-conductors, where the current carrying species is the electron, whereas types III and IV give rise the p-type semiconductors, the conduction being due to the existence of positive holes.

At temperatures above the Tammann temperature (~0.5 T_m, where T_m is the melting point/K) [23] both ions and electrons are mobile, and there is ion movement between interstitial sites, and from a site into the neighbouring vacancy. Thus at such temperatures there is a rise in conductance due to this ionic mobility.

4.3.3 Impurity ions

The addition of a foreign ion to a lattice position in a crystal always causes a degree of disorder even if it has the same charge. Since similar charged ions of different metals for example, differ normally in size, polarizability, etc., then they will have different interaction energies in the lattice. Thus substituting Fe^{3+} for Al^{3+} in an Al_2O_3 crystal will necessitate some degree of relaxation of the crystal to accommodate it, and the same will be true if the substitution occurs at the surface. There is also some evidence that defect concentrations of this type are greater near the surface. If the foreign ion valency is different from that of the normal lattice ion, then the situation is very similar to that pertaining in non-stoichiometric crystals. Fig. 4.12 illustrates the addition of the oxides L_2O and R_2O_3 to the divalent oxide MO, where L and R are mono- and trivalent metal ions respectively, producing anion and cation vacancies respectively. If the

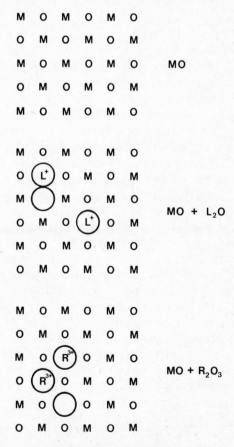

Fig. 4.12 – The addition of the monovalent metal oxide L_2O and the trivalent metal oxide R_2O_3 to the divalent metal oxide, MO.

metal M can exist in a higher oxidation state, say M^{3+}, then in the case where L_2O was added the oxide ion vacancy could be filled and the condition of electrical neutrality maintained if two M^{2+} ions became M^{3+}. Analogously in the case of addition of R_2O_3, the metal ion vacancy could be filled by an M^{2+} ion if two M^{2+} ions became M^+ ions, or if there existed in the crystal two M^{3+} ions, they reverted to M^{2+} ions. Therefore the substitution of lower valent metal ions would result in an increase in the number of positive holes, and higher valent metal ions a decrease.

The introduction of n-type semiconduction by isomorphous replacement is possible. Barium titanate exists in the perovskite structure, and it is possible to replace some of the Ba^{2+} ($r = 0.135$ nm) by La^{3+} ($r = 0.115$ nm), electrical neutrality being maintained by the conversion of Ti^{4+} to Ti^{3+}. Thus the resulting structure may be represented as $La_x^{3+} Ba_{1-x}^{2+} Ti_x^{3+} Ti_{1-x}^{4+} O_3$ and n-type semiconduction is the result of Ti^{3+} ions on Ti^{4+} ion sites.

4.3.4 Dislocations

Although it is not a precise definition [24], dislocations may be thought of as concentrations of point defects in the crystal lattice. They are the result of the processes of nucleation, crystal growth, and if the solid was precipitated from solution, may be flocculation as well. The two most significant types are known as edge and screw dislocations. Cottrell diagrammatically represented an edge dislocation in say a simple cubic rare gas crystal lattice as shown in Fig. 4.13(a), where the broken line AB represents the slip plane, perpendicular to the plane of the figure. The lattice extensions above and below the figure can be regarded as normal cubic crystal lattice, but in the upper part of the figure lattice spacings are being expanded, and in the lower half compressed, to accommodate the jump in the number of atoms in a row at slip plane. Such dislocations, the dislocation being denoted by the symbol T in Fig. 4.13(a), emerge at the surface as a step with higher local surface energy, or external pressure may result in the restoration of registration. Their presence is characterised by the possibility of relative movement in the plane of the dislocation when only small stresses are applied. The gliding of such dislocations is postulated as the mechanism of plastic flow, and the process has been likened by Adamson [25] to "moving a rug by pushing a crease down it". It is possible to quantify the nature of such a dislocation in terms of a *Burgers circuit* [26]. This is any site to site path taken in the 'good' part of a crystal containing dislocations which forms a closed loop. If the same path sequence in a perfect crystal does not form a closed loop, then the circuit in the first crystal must contain one or more dislocations; the vector required to close the loop is Burgers vector. This is illustrated in Fig. 4.14 where the circuit MNOPQ in (a) does not close in (b), and the closure failure QM is the Burgers vector, which is perpendicular to the dislocation line, that is, the perpendicular at T in Fig. 4.13(a).

The other common form of dislocation, the screw dislocation [27], is illustrated in Fig. 4.13(b) (where each cube represents a lattice site or atom), and is characterised by a Burgers vector parallel to the line of the dislocation, the perpendicular at A. Looking down the line of dislocation, a fall of one layer takes place in making a 360° turn from B about A, until the line BC is met again. Since this motion is anticlockwise, the screw dislocation in the figure is termed left-handed.

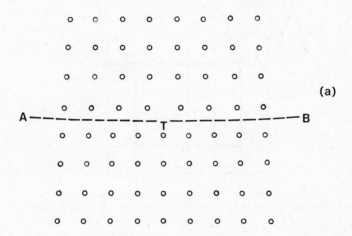

(a)

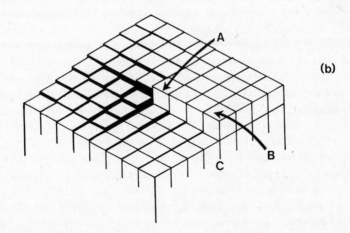

(b)

Fig. 4.13 (a) – Cottrell's [24] representation of an edge dislocation, and
(b) – Frank's [27] representation of a screw dislocation.

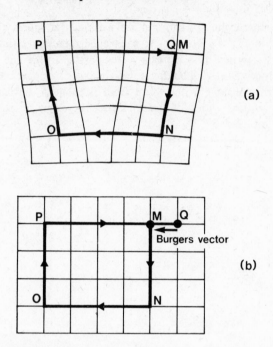

Fig. 4.14 (a) – A Burgers circuit around a dislocation, and
(b) – the same circuit in a perfect crystal; the closure failure is Burgers vector.

4.3.5 Crystal growth

The idea that the steps on the surface produced by dislocations were responsible
for the growth of a crystal face was due to Frank [27], which together with
Burton and Cabrera's [28] analysis of the atomic nature of a growing surface,
explained [29] the experimentally observable high growth rates on densely
packed faces at low supersaturations and predicted that spirals should be visible
on the crystal surface. Experimental verification soon followed when Griffin [30]
observed them on natural beryl crystals using a metallurgical microscope, Verma
and Amelinckx [31(a)] found them on carborundum, and they were also seen
on cadmium iodide at about the same time [32]. Dawson and Vand [33] also
found them on paraffin crystals using electron microscopy.[†]

The spirals that have been observed are one cell constant in height at each
step. Thus we can think of growth as the adsorption of fresh material at the step
face causing each face to develop by the movement of a portion from the
centre of a face to the edge. That edge would grow in preference to a face, and
that the edge would grow uniformly, can be thought of as a consequence of the
adsorption energy being greatest where the number of possible nearest neighbours
is greatest, thus an edge would be preferred to a face and a kink site on an edge

[†] A collection of some of the micrographs taken at this time can be found in Ref. [24(c)], p.150.

above all. This process is illustrated in Fig. 4.15, which is similar in many respects to Fig. 4.13(b), except that distortion of the structural cubes has been permitted. Growth can continue at such an edge without the need of further nucleation.

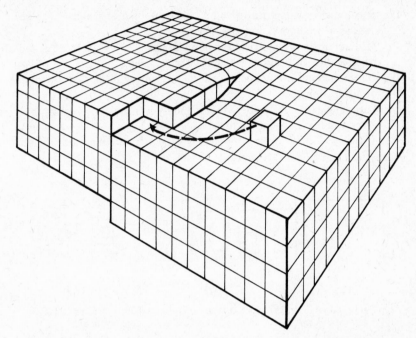

Fig. 4.15 – The preferred position for addition to a surface screw dislocation. The upper surface is a spiral ramp and thus the step cannot be removed by adding further atoms to it.

The density of dislocations is usually stated in terms of the number of dislocation lines intersecting unit area in the crystal, and it usually quoted as about 10^{12} m^{-2} for 'good' crystals[†] up to about 10^{16} m^{-2} for cold worked metals. This suggests dislocations would be separated by about 10-10^{3} nm, or in other words every crystal larger than about 10 nm will have dislocations in its surface, and one surface atom in every thousand is likely to be adjacent to a dislocation.

The structure around a dislocation line is severely strained, and as a result the chemical potential of the material in such locations will be increased. If this happens to a large enough extent a hollow may appear [34], but it is only very probable in crystals with large unit cells and therefore correspondingly large Burgers vectors, it is somewhat surprising therefore that they were observed first in carborundum [31(b)].

[†] In Griffin's [30] beryl crystals the maximum value was only ~10^{10} m^{-2}.

Dame Kathleen Lonsdale [35] pointed out that most real crystals are mosaics in which component crystallites may or may not be well aligned, as illustrated in Fig. 4.16. Four main types of grain boundary may be distinguished [36]:

I. Small angle symmetrical tilt ($\theta < 1°$) composed of parallel edge dislocations which have not contracted.

II. Medium angle symmetrical tilt boundaries ($5° < \theta < 20°$) composed of parallel edge dislocations which have contracted.

III. Unsymmetrical boundaries ($\theta < 20°$) involving more than one family of dislocations.

IV. Large angle ($\theta \simeq 30°$) where the boundary is incoherent, that is, there is no continuity along lattice rows.

The commonest of these crystal mosaics is type III, and they are immobile, whereas types I and IV are very mobile.

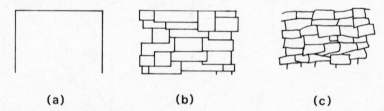

(a)　　　　　　　(b)　　　　　　　(c)

Fig. 4.16 – Lonsdale's [35] representation of (a) a perfect crystal, (b) a mosaic of parallel crystallites, and (c) a mosaic of disordered crystallites.

4.3.6 Surface roughness

This particular term often has a somewhat ill-defined meaning [37]. To quote [38]:

"When *I* use a word," Humpty Dumpty said, in rather a scornful tone, "it means just what I choose it to mean — neither more nor less".

"The question is," said Alice, "whether you *can* make words mean as many different things."

"The question is," said Humpty Dumpty, "which is to be master — that's all." Humpty Dumpty's comments are frequently apt in so many walks of life, but particularly so in this case, since what is rough and what is smooth depends on the scale of the observation of the surface. For example, riding along a *smooth* road on a bicycle, until, on falling off, one is grazed by the *roughness* of the surface: one's point of view is altered by the closeness of the observation.

There is a level of variation at which the term 'roughness' appears inappropriate, that is when we look at positional variations caused by the vibration of molecules or atoms about lattice positions. For example, the vibrational amplitude of palladium atoms in the surface of the metal is about 0.014 nm, as measured with low energy electrons [39], compared with about half that value in the bulk of the metal. In general 'roughness' is confined to changes in the surface representing movements of the surface larger than the interatomic distances. Suppose a cubic crystal of say sodium chloride was cleaved parallel to the face, then the exposed surface might well look like that shown in Fig. 4.17(a), in which all the faces exposed are identical cubic faces. However, if such a crystal were to be cut at a small angle to this plane, the resulting surface is then a ladder, as shown in Fig. 4.17(b). Although the upper face of each step will be the same cubic face as in Fig. 4.17(a), the transition from step to step may be achieved using the same crystal face (1) or may involve a different crystal plane (2). Since the packing density varies between different crystal planes, their relaxed states at a surface would normally imply different surface energies. A single crystal of uranium (IV) oxide cleaved at an angle of 11° to the (111) plane was shown by low energy electron diffraction patterns [40] to exhibit a 'ladder' type of surface.

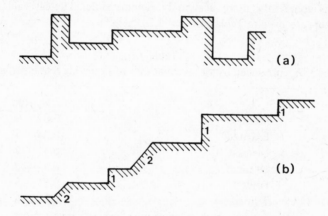

Fig. 4.17 (a) – The profile of a cleavage face parallel to a cube face,
(b) – The profile of a cleavage face cut at a small angle to the cube face.

Most solid surfaces that one meets are in fact prepared in some way: sawn, cut, turned, polished or chemically treated, for example. All of these methods leave the surface rough, if we adopt the definition of the previous paragraph, to differing degrees. To quantify estimates of roughness one may make use of two parameters, one related to the amplitude of the vagueness of the surface, and the other to the increase in area associated with the roughness. In Fig. 4.18, the real surface is XY and AB represents the surface of a hypothetical solid of equal volume with a molecularly smooth surface. Thus it is located in the position where if the 'hills' in XY above the plane AB were removed they would be equal in volume to the 'valleys' in XY below the plane AB. The amplitude of the variations in the surface may be expressed as

$$h_{av} = \frac{|h_1| + |h_2| + |h_3| + |h_4| + \ldots |h_n|}{n} \tag{4.21}$$

where h is the distance of the profile XY from AB measured at a large number of points on the surface, or

$$h_{rms} = \sqrt{\frac{h_1^2 + h_2^2 + h_3^2 + h_4^2 + \ldots h_n^2}{n}} \tag{4.22}$$

The roughness factor, r, is defined by:

$$r = \frac{\text{Area of the real surface XY}}{\text{Area of the surface defined by AB}} \tag{4.23}$$

which is frequently more accurately determinable. Typical values of these parameters are given in Tables 4.1 and 4.2.

Table 4.1
Values of h_{av} produced by various methods of preparing metal surfaces [41].

	$h_{av}/\mu m$
Polished	$0.02 - 0.25$
Extruded	$0.25 - 4$
Die cast	$0.4 - 4$
Ground	$0.5 - 2.5$
Drilled	$2.5 - 5.$
Turned	$3 - 6$

Table 4.2
Measurements of the roughness factor, r, for various surfaces.

	r	Method	ref.
Glass beads, once cleaned	1.6	Gas adsorption	[42]
Glass beads, twice cleaned	2.2		
Glass beads, thoroughly cleaned	5.4	Dye adsorption	[43]
Silver foil	5	Double layer capacity	[44]
Silver foil, etched	15		
Steel, electropolished	1.12	Gas adsorption	[45]

Fig. 4.18 – A profile of a surface of a solid (X-Y). The line A-B locates the surface of a hypothetical solid of equal volume. The distance h_{max} is the distance between the lowest and highest points on XY.

Surfaces are almost always contaminated with foreign material unless extra care is taken specifically to clean them and keep them clean. Many metals normally have an oxidised surface, and the layer of oxide present is often in the range 1-10 nm thick. Metal surfaces therefore often behave like the oxide and not like the metal beneath. As an example which well illustrates the sort of complexities which might be encountered, we may look at the aluminium surface [46]. In dry air the oxide thickness is 1-3 nm, but if moisture is present the layer is thicker, >10 nm, and grows almost indefinitely. In the latter case it is usually stratified, a dense, amorphous, water-free layer on the metal, and above it a thick, porous hydrated region next to the air.

The polishing process also illustrates the role past history can play in determining the nature of the surface. The surface layer produced by polishing is usually known as a Beilby layer and appears to be amorphous under the microscope [47], having the appearance of liquid viscous film which has flowed into the irregularities of the surface [48]. Electron diffraction studies of this layer show diffuse rings typical of the amorphous state, but Raether [49] has concluded that it is in fact microcrystalline, with crystallites so small that they give rise to diffuse electron diffraction rings. A typical cross-section of a polished surface is shown in Fig. 4.19, and illustrates that polishing also produces considerable deformation of the underlying metal.

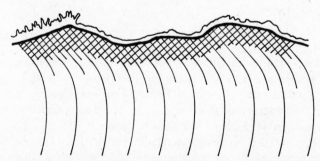

Fig. 4.19 – Samuels' model of a polished surface. ∿∿ oxide ca 0.01 – 0.1μm, ➤ Beilby layer 0.1μm, ✸ severe deformation 1 – 2μm, ＼ gross deformation 5 – 10μm, ＼＼ minor deformation 20 – 50μm (After Bowden and Tabor [50]).

There is also evidence of oxide being dragged into the Beilby layer [51] as well as contamination by particles of the polishing medium [52], furthermore there is evidence that the Beilby layer is unstable, recrystallisation to larger crystallites occurring [53]. Electropolished surfaces are usually less extreme than polished ones, usually showing crystalline diffraction patterns [54].

4.3.7 Heterogeneity

As the reader will readily appreciate, all the previous sub-sections of this section deal with factors which would cause a surface to be heterogeneous, which is another Humpty Dumpty word [38] (see 4.3.6). But before seeking to define it more closely, let us consider the fact that as perfect a simple cubic sodium chloride crystal as one could reasonably conceive, somehow free of all the defects etc. we have previously considered (an impossibility since it could not have grown, although miraculous cleavage of a larger crystal might produce it with divine intervention), would still be heterogeneous. In section 4.2.2 we discussed the calculation of surface energies by a summation technique over a semi-infinite hemi-crystal, and it can be shown that quite a large number of nearest neighbours need to be considered if the forces on one ion in the surface are to be calculated, say of the order of a thousand. This could be represented by a hemisphere some 17 unit cell distances in radius for NaCl. Thus the energy of any ion within 17 ions of the edge of the crystal will have a different energy from at the centre of the face. In general there will be an increase in the free energy of the ions as the edge is approached, and obviously one would predict even greater effects as a corner is approached. However, the exact magnitude of the problem is uncertain because the effects of relaxation of the crystal will likewise change from the behaviour in the centre of face. An early calculation of Benson and Schreiber [55] produced a value of 29 pJ m^{-1} for an edge, when no allowance was made for distortion.

Even for an ideal surface without edges or corners, the local value of U^s would differ significantly from $\overline{U}^s$, the average value for the surface. This could be expressed statistically as

$$\overline{U}^s = \int_0^\infty f(U^s)\mathrm{d}U^s \tag{4.24}$$

where $f(U^s)$ is a distribution function, but to get a really detailed picture of the surface we would also need a correlation function giving the distribution of local values of U^s as a function of position. Because a surface of sodium and chloride ions cannot be homogeneous (chloride ions are not the same as sodium ions) Ross proposed that a new word should be coined to denote as perfect a surface as an infinite crystal plane free from defects could achieve, and that is *homotattic* [2]. It might also be reasonable to talk of a plane of all one type of atom as *homogeneous* or *uniform*, such as an ideal graphite plane.

Thus all surfaces which are not *homogeneous*, *uniform* or *homotattic*, may be described as *heterogeneous*.

4.4 SURFACE MOBILITY

The impression may have been given in this chapter that the surface of a solid is a permanent structure, which unless treated in some way, will remain immobile. This need not necessarily be the case, but in order to put some perspective on the situation, let us first examine the availability of molecules at a water-vapour interface in terms of the evaporation-condensation equilibrium at say 25°C. The kinetic theory of gases predicts that the number of molecules striking unit area in unit time, n', will be given by

$$n' = P/(2\pi m k T)^{\frac{1}{2}} \qquad (4.25)$$

where m is the mass of one molecule, and the other symbols have their usual significance. The vapour pressure of water is 3.2 kPa at this temperature, which leads to a value of 1.1×10^{26} molec $m^{-2} s^{-1}$ for n'. From the water density, 1 m^2 of surface should contain 10^{10} molecules, therefore if we assume all collisions with the surface are effective and that there is a balance between evaporation and condensation rates, then the lifetime of molecule in the water surface will be $\sim 10^{-7}$ s. In contrast a similar calculation for solid tungsten using the value of 10^{-35} Pa for the vapour pressure gives a value of the surface lifetime of $\sim 4 \times 10^{24}$ years, and for ice at 0°C a lifetime of $\sim 5 \times 10^{-7}$ s. These values for solids are likely to be much too small, since it is highly unlikely that every collision will result in capture. In fact for liquid water estimates as low as 0.034 for the fraction of collisions that will be effective have been postulated [56], and this would change the value for the lifetime in the liquid water surface at 25°C to 2.7 μs. Therefore although we cannot get a precise estimate of lifetimes in solid surfaces, the range is likely to be enormous; from the permanency of the tungsten surface to the extreme mobility of ice.

There is quite persuasive evidence to support this idea that the mobility of solid surfaces is appreciable near the melting point of the solid, and much less at lower temperatures. Small ice spheres at −10°C tend to coalesce at the points of contact, and a careful study of the kinetics of the process [57,58] permits the identification of whether bulk or surface diffusion is the most important. Scratches on a silver surface also fill up and disappear when it is heated to near the melting point of silver [59], and if a silver sphere is placed on a silver surface at the same temperature, there is appreciable flow at the point of contact, making the junction unidentifiable [60]. There is also the observation that a thin copper wire near its melting point tends to shorten, and from the stress required to reduce the strain to zero, a surface tension value of 1.37 N m^{-1} was calculated [61] (much greater than most liquid values). Such a measurement could not have been made if the system was not appreciably mobile.

Both surface and bulk mobility are obviously involved in the process of sintering. In this process, a powdered sample becomes welded into a solid matrix at the point of contact between the particles, and is accompanied by a reduction in surface area. This process is important in the range between the Tammann temperature and the melting point, and the Tammann temperature was defined empirically [62] as that temperature above which the rate of sintering increased markedly. Often it may be taken as $0.5\ T_m$, but in fact usually values are in the range $0.37 - 0.5\ T_m$. Its physical significance is thought to be that bulk crystalline defects become mobile [63], thus allowing the transport of material. The analogous mobility of surface defects starts at a lower temperature, as would be expected, usually near $0.3\ T_m$ [64,65]. Further details can be found in the reviews by Herring [66] and Kuczynski [67].

4.5 SURFACE GROUPS

Because the solid has a particular bulk chemical composition, it does not necessarily follow that the surface contains the same atoms arranged in a similar way to the bulk phase. Often as the result of chemical interactions with the components of the surrounding gas phase, a number of different chemical groups may be found on the surface. The best understood examples of this are metallic oxides [68], silica [69] and carbons [70].

As an example of the behaviour of metal oxides, rutile presents a typical case. The changes in the rutile spectra caused by heating and evacuation were explained by Jackson and Parfitt [71] in terms of the bridged and terminal hydroxyl groups formed by the dissociation of adsorbed water on the (110) plane.

Hockey and Jones [72] found differences in the behaviour of (100), (101) and (110) planes, suggesting that the first two absorb molecular water as a ligand co-ordinated to a surface Ti^{4+} ion, whereas the adsorption of water on the (110) plane involves the dissociation of the water molecule and the formation of two differently bonded OH^- ions. Calculations by Waldsax and Jaycock [73,74] suggest that the water molecule would adsorb preferentially on the five co-ordinate Ti^{4+} sites in this plane, and then dissociate to give the two types of hydroxyl group.

A silica particle can be regarded as a polymer of silicic acid [75], which terminates at the surface in either a siloxane grouping, $\geq$Si$-$O$-$Si$\leq$, with the oxygen in the surface, or as the result of interaction with water, one of several forms of silanol group, $\geq$Si$-$OH. These can be defined as follows:

isolated $-$ where the surface Si atom has one bond attached to the $-$OH group which is not hydrogen bonded to another silanol group;

vicinal $-$ where two single silanol groups attached to different Si atoms are close enough to hydrogen bond to each other; and

geminal $-$ where two $-$OH groups attached to the same Si atom, leaving only two bonds from the Si atom to the bulk structure, which are too close to hydrogen bond to each other.

Commercial carbon blacks are usually found to possess five major chemical groups on their surface [76-78]; aromatic hydrogen, phenol, quinone, carboxylic acid and lactone. Rivin [79] found for a furnace black with a surface area of 76 m^2g^{-1} the following concentration of surface groups, expressed in milli-equivalents g^{-1}:

aromatic hydrogen	1.94	(0.065)
phenol	0.61	(0.21)
quinone	0.00	
carboxylic acid	0.01	(12)
lactone	0.17	(0.74)

where the figures in brackets represent the area of surface per group in nm^2. It is possible that some of the hydrogen may not in fact be in the surface but may penetrate the bulk of the carbon.

The discussion in this section should underline, if such underlining is needed, the danger of assuming that the surface of a solid is closely related to the bulk structure. Such an assumption may occasionally be a reasonable approximation, but often can be inaccurate and misleading.

4.6 THE ADSORPTION OF GASES AND VAPOURS

There are a number of ways of defining the process of adsorption, but one of the more illuminating is in terms of degrees of freedom. As a molecule or atom approaches a solid surface, the intermolecular attractive forces between the approaching molecule (the adsorptive) and the atoms, ions or molecules comprising the solid (the adsorbent), results in a net attractive force on the molecule drawing it towards the surface, and a decrease in the internal energy of the system as a whole. It is the relative magnitudes of the translational kinetic energy of the gas molecule and the adsorption energy, that determine whether the molecule is adsorbed. In general we may talk of a molecule or atom as being adsorbed (the adsorbate) if it has lost at least one degree, and maybe all three, of translational freedom.

4.6.1 Physical and chemical adsorption

The distinction between these two types of adsorption lies in the nature of the forces causing the loss of translational freedom. If these are essentially London-van der Waals forces and electric field-dipole interactions, then the energies involved are usually relatively low, that is less than about 40 kJ mol^{-1}. Chemical bond formation on the other hand results in the much larger energies normally associated with chemical bonds.

The chemical adsorption process of necessity first involves a physical adsorption process as the molecule approaches the surface, because of the longer range of the physical forces, and the transition from physical to chemisorption, may or may not be associated with an activation energy barrier. This process was first visualised by Lennard-Jones [80] and is frequently reproduced diagrammatically [81], and is shown in Fig. 4.20. Here the molecule A_2 approaching say a metal surface, would have potential energy represented by the curve labelled $M + A_2$. There is a shallow minimum of depth U_P, close to the distance corresponding to the sum of the van der Waals radii of the approaching molecule and the surface atom. The curve labelled $M + A + A$ represents the potential energy of two dissociated atoms of A, and at a distance far from the surface is separated from the first curve by the dissociation energy of A_2. It is the chemical bond formation of two M-A bonds that gives the deeper minimum U_C. Since the two curves cross at a point P, then as a molecule approaches the surface it can dissociate and cross from the $M + A_2$ curve to the $M + A + A$ curve. Because of wave mechanical effects the transition in fact follows the dashed line in the vicinity of P. Thus when the relationship between the curves is as shown in Fig. 4.20(a), chemisorption should occur easily. The barrier to the changeover would be small or non-existent, since the vibrational energy level in the minimum U_P, may be higher than the dashed curve.

In Fig. 4.20(b), however, we have the case where there is an appreciable activation energy barrier, and E is a significant fraction of the dissociation energy. In this case chemisorption may become slow at low temperatures although it corresponds to a thermodynamically more favourable state.

This type of dissociative chemisorption is responsible for the formation of oxide films on metals and the formation of hydride films on metals when hydrogen is adsorbed. The adsorption of water on oxides is also often a case of dissociative chemisorption where for example

$$M = O + H_2O \longrightarrow M \begin{cases} OH \\ OH \end{cases}$$

giving an hydroxylated surface which subsequently physically adsorbs water, if we assume hydrogen bonding is a physical process. As an example of this we might consider the hydroxylation of a (110) rutile surface, since on this surface the value of U_P has been calculated [74] to be 434 kJ mol^{-1}; the first dissociation energy of water is 477 kJ mol^{-1}; and the TiO bond energy is 657 kJ mol^{-1}, thus it would seem probable that the transition from the U_P to U_C minimum would correspond, in essence, to Fig. 4.20(a).

It has to be admitted that this picture of the dissociative chemisorption process is too simple, and that there probably will be intermediate states between the two curves in many cases, but nevertheless the idea of physical adsorption followed by rearrangement, bond rupture and chemisorption is useful.

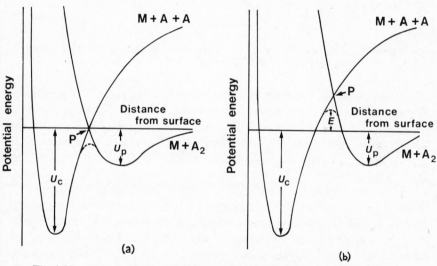

Fig. 4.20 – The chemisorption process involving the dissociation of molecule A_2 and the formation of two M-A bonds.

4.6.2 Relation between adsorption potential and the heats of adsorption

The potential energy curve for adsorbate-adsorbent interaction, as illustrated in Fig. 4.21 is the result of the sum of short-range Born repulsive forces and longer range van der Waals attractive forces, but the relationship between potential energy curves and experimentally determined heats of adsorption is not usually that clear, for reasons that can be best explained with the aid of the figure.

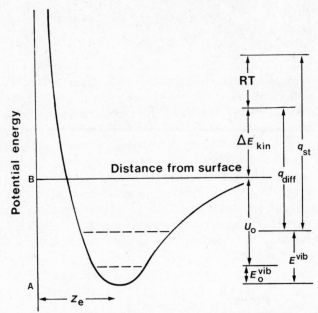

Fig. 4.21 – The potential energy of interaction of an adsorbate molecule with the adsorbent, and the relationship to the heats of adsorption.

Firstly there are two ways of fixing the zero of the potential energy scale. The minimum value in the potential energy well, level 'A' in the figure, may be taken as zero [82], or the level 'B', that is, the potential energy when the adsorbate molecule is a large distance away from the surface, may be chosen as zero: the latter is a common choice when potential energy calculations are made. The relationships between the various quantities remain unaltered by this choice.

It is not possible for a molecule to remain stationary at the distance z_e, at the bottom of the potential energy well. Even at 0 K it will possess vibrational energy, and E_0^{vib} is the vibrational energy at this temperature. When the temperature is increased it is possible to excite it to a higher vibrational level, and at the temperature T, this is shown in the figure as E^{vib}. Thus the problem in relating the differential heat of adsorption, q_{diff}, and the isosteric heat of adsorption, q_{st}, to the adsorption potential, U_0, is that usually the vibrational frequencies (and therefore E_0^{vib} and E^{vib}) are unknown.

The isosteric heat of adsorption is determined from a family of adsorption isotherms at various temperatures, since

$$(\partial(\ln P)/\partial T)_{n_a} = (\partial(\ln P)/\partial T)_\theta = q_{st}/RT^2 \qquad (4.26)$$

where P is the equilibrium gas phase pressure, and n_a is the amount adsorbed. If the temperature range is small, then constant n_a is equivalent to constant fractional coverage, θ. This equation can be used to derive q_{st} from the plot of

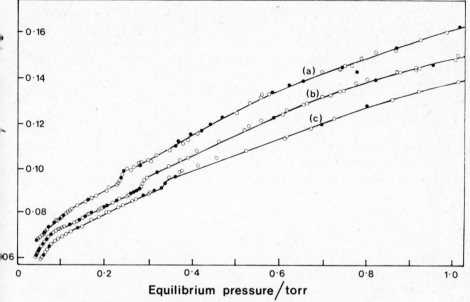

Fig. 4.22 – Adsorption isotherms for krypton adsorbed on silver iodide at (a) 77.49, (b) 78.30 and (c) 79.10 K in the pressure range 0.04 – 1.00 torr (1 torr = 133.3 Pa). (From Sidebottom *et al.* [83]).

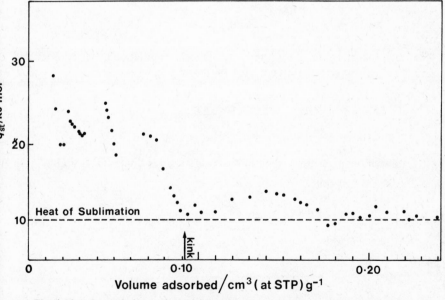

Fig. 4.23 – Isosteric heat values as a function of coverage from the data shown in part in Fig. 4.22.

$\ln P$ vs $1/T$, since the slope is equal to $-q_{st}/R$. Since q_{st} is not entirely independent of T, then a number of accurate experimental isotherms are necessary to evaluate q_{st}. Adsorption isotherms [83] for krypton on a silver iodide sample at temperatures $\sim$78 K are shown in part in Fig. 4.22 and the derived isosteric heat values in Fig. 4.23.

4.6.3 Residence times

Since the equilibrium between gaseous and adsorbed phases must represent a balance between adsorption and desorption rates, then there must be a finite residence time for an adsorbed molecule on the surface. In 1924 Frenkel [84] proposed that this could be expressed as

$$\tau = \tau_o \exp\left(q_{st}/RT\right) \tag{4.27}$$

where τ_o is the oscillation time for vibration of the adsorbed molecule perpendicular to the surface. However, the problem is perhaps best treated by statistical thermodynamic means [85, 86].

At adsorption equilibrium for a single gas, the chemical potentials of the adsorbed and gaseous phases must be equal. If the number of molecules in the adsorbed and gaseous states are N_a and N_g respectively, and $_af$ and $_gf$ are the corresponding partition functions, then

$$\frac{_gf}{N_g} = \frac{_af}{N_a} \exp\left(q_{st}/RT\right) \quad . \tag{4.28}$$

For each partition function there are possible translational, vibrational and rotational contributions, thus equation (4.28) may be written

$$\frac{_gf_{tr} \cdot _gf_{vib} \cdot _gf_{rot}}{N_g} = \frac{_af_{tr} \cdot _af_{vib} \cdot _af_{rot}}{N_a} \exp\left(q_{st}/RT\right) \quad . \tag{4.29}$$

The translational partition functions for ideal three-dimensional and two dimensional gases are given by

$$_gf_{tr} = \left[\frac{2\pi m\, kT}{h^2}\right]^{3/2} \frac{N_g\, kT}{P} \tag{4.30}$$

and

$$_af_{tr} = \frac{2\pi m\, kT}{h^2}\, S \tag{4.31}$$

where S is the surface area. If, however, translation is hindered in the adsorbed state then

$$_a f_{tr} = \frac{2\pi m\, kT}{h^2}\, S \times \frac{_a f_{tr}\ \text{(hindered)}}{_a f_{tr}\ \text{(ideal 2d)}} \quad . \tag{4.32}$$

If adsorption does not alter the internal vibrations of the molecule, but converts one degree of translational freedom into a vibrational mode perpendicular to the surface, then

$$_a f_{vib} = {}_g f_{vib}\cdot f_z \tag{4.33}$$

where f_z is the vibrational partition function for the additional vibration. Combining equations (4.29) to (4.33) gives

$$\frac{N_a}{S} = \frac{P}{(2\pi m\, kT)^{1/2}}\, \frac{h}{kT}\cdot f_z\cdot \frac{_a f_{tr}\ \text{(hindered)}}{_a f_{tr}\ \text{(ideal 2d)}} \cdot \frac{_a f_{rot}}{_g f_{rot}} \exp\,(q_{st}/RT) \quad . \tag{4.34}$$

Comparing this equation with equation (4.27) indicates that

$$\tau_0 = \frac{h}{kT}\cdot f_z\cdot \frac{_a f_{tr}\ \text{(hindered)}}{_a f_{tr}\ \text{(ideal 2d)}} \cdot \frac{_a f_{rot}}{_g f_{rot}} \quad . \tag{4.35}$$

Thus for the case where rotation of the molecule is unrestricted on adsorption (for example, a monatomic gas) and the film is truly mobile then

$$\tau_0 = \frac{h}{kT}\, f_z \quad . \tag{4.36}$$

It is possible therefore to calculate a value for τ_0 if a value for f_z can be determined. If the vibration perpendicular to the surface *is* excited, the frequency is low, and the partition function will be given by $f_z = kT/h\nu_z$, then $\tau_0 = 1/\nu_z$, which would be consistent with Frenkel's original ideas. However, if the vibrational mode is unexcited, then $f_z = 1$ and

$$\tau_0 = h/kT \quad . \tag{4.37}$$

If translation on the surface is hindered then equation (4.35) would indicate a lower value of τ_0, and in the extreme case of localised adsorption the derivation would have to be modified to account for entropy changes associated with distributing N_a molecules over N_s adsorption sites. In this case, if the perpendicular vibrational mode is again unexcited, then

$$\tau_0 = \frac{h}{kT} \cdot \frac{h^2 N_s / S}{2\pi m\, kT} \cdot \frac{{}_a f_{rot}}{{}_g f_{rot}} \quad . \tag{4.38}$$

De Boer has suggested that for the localised adsorption of water on charcoal at 300 K, the value of τ_0 is $\sim 10^{-16}$ s.

To assess the implications of this approach on residence time, it is pertinent to calculate τ at various temperatures, but to do this we need some assumptions about the change in τ_0 for a localised film with temperature. If the ratio of rotational partition functions is unaltered, then equation (4.38) indicates that $\tau_0 \ \alpha T^{-2}$. Thus at 77.5 K the value of τ_0 for a localised film is $\sim 1.5 \times 10^{-15}$ s.

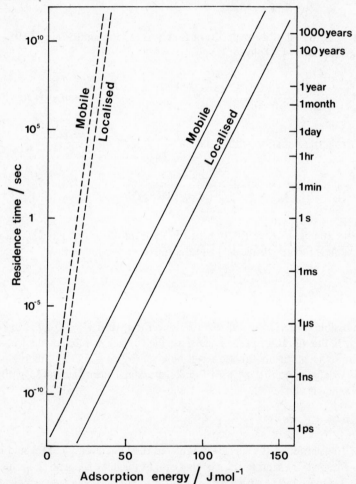

Fig. 4.24 – Approximate residence times of molecules in mobile and localised adsorbed films as a function of adsorption energy, at 298.15 K ——— and 77.5 K - - - -.

Fig. 4.24 shows approximate values of τ calculated using the value of τ_0 from equation (4.37) for mobile films (1.6×10^{-13} s and 6.2×10^{-13} s at 298.15 K and 77.5 K respectively, and the values quoted in the preceding paragraph for localised films.

4.6.4 Adsorption potential variations on a real surface

It is important to realise that the adsorption potential of even such a simple adsorbate as a monatomic gas, will vary with position on the surface of even such a surface as an ideal graphite adsorbent. If we define a quantity $_m U_0$ given by

$$_m U_0 = U_0 + E_0^{vib} \tag{4.39}$$

then this could be evaluated by calculating all the separate interactions between the adsorbed atom at a particular position at distance d from each of the carbon atoms in the substrate, and then summing the contributions for each individual interaction. Thus the interaction of the gas molecule (i) with each of the atoms (j) of the adsorbent could be written

$$u_{ij} = C_1 d_{ij}^{-6} + C_2 d_{ij}^{-8} + C_3 d_{ij}^{-10} + R d_{ij}^{-12} \tag{4.40}$$

where C_1, C_2 and C_3 are the dipole-dipole, dipole-quadrupole and quadrupole-quadrupole coefficients, respectively; R is the Born repulsion constant. The total interaction will therefore be

$$_m U_0 = \sum_j u_{ij} = C_1 \sum_j d_{ij}^{-6} + C_2 \sum_j d_{ij}^{-8} + C_3 \sum_j d_{ij}^{-10} + R \sum_j d_{ij}^{-12} \ . \tag{4.41}$$

Avgul *et al* [87] have calculated values of $_m U_0$ for neon, argon and krypton at three different positions on a graphite surface, and these are illustrated in Fig. 4.25. Their results are tabulated in Table 4.3.

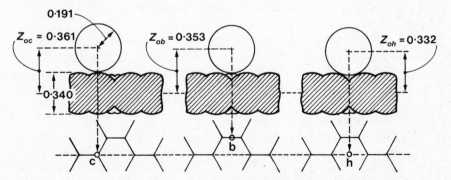

Fig. 4.25 – Adsorbed argon atoms at different positions on a graphite substrate; all distances in nm (From Avgul *et al* [87]).

Table 4.3

Calculated values of $_mU_0$/kJ mol^{-1} for inert gas molecules at three positions on a graphite surface [87].

	c above C atom	b between 2 C atoms	h centre of C ring	Average
Neon	3.22	3.51	4.64	3.81
Argon	8.16	8.66	11.05	9.29
Krypton	10.84	11.63	14.52	12.34

This table clearly illustrates two particularly significant facts. Firstly the increase in $_mU_0$ with increasing adsorbate molecular size, and secondly the appreciable variation in $_mU_0$ with position. Whether these variations would be large enough to hinder the mobility of an adsorbed inert gas atom would be dependent on the relationship of the variation in $_mU_0$ to the energy associated with one translational degree of freedom, $\frac{1}{2}kT$. At 77.5 K, the latter corresponds to 0.16 kJ mol^{-1}, and the variation in $_mU_0$ from the table is 1.42 kJ mol^{-1} in the case of neon and 3.68 kJ mol^{-1} for krypton. Direct comparison however takes no account of the magnitude of E_0^{vib}: we will return to this point again later.

In the case we have just considered there is a high degree of symmetry in the surface, and the positional variations can be easily represented in the table. Also neglected in the calculation were the effects of relaxation of the surface from bulk lattice positions as the surface is approached. In this particular case, since only dispersion interactions are involved and because of the nature of graphite, the bulk of the contributions to $_mU_0$ come from the nearest neighbours, and also probably relaxation of the graphite surface involves the movement of whole lattice planes. Therefore the overall picture described in Table 4.3 will be substantially correct.

The calculation also excluded the possibility of secondary relaxation of the surface brought about by the adsorption of the inert gas atom itself. In this case again the effect is unlikely to be significant since $_mU_0$ is much smaller than the bond energy holding the graphite plane together. However, if the forces holding the solid together were comparable with the adsorption forces, then the effect might be significant.

Burton and Jura [17] considered the effect of relaxation on the value of $_mU_0$ for argon, neon and krypton adsorbed on the surface of solid argon, and their results are shown in Table 4.4. The changes are clearly significant in terms of the value of $_mU_0$, and there are also positional changes as shown in Fig. 4.8 for the adsorption of an additional argon atom. The nearest neighbours to the adsorbed atom relax towards the bulk 0.6% of the lattice spacing from their normal surface positions and outwards tangentially to the surface by 0.3%, as previously indicated in section 4.3.1.

Table 4.4

Values of $_mU_0$/kJ mol^{-1} for argon, neon and krypton adsorbed on the (100) face of solid argon, with and without allowance for relaxation [17].

	Argon	Neon	Krypton
Without relaxation	5.67	2.48	6.60
With relaxation	5.72	2.87	7.04

4.6.4.1 Adsorption potential maps: solid inert gases

The problem of calculating adsorption potentials before about 1960, was largely the amount of tedious arithmetic that had to be done. The advent of the modern computer has changed the situation drastically. Although the work involved in writing an adequate program and debugging it is not inconsiderable, the additional work involved in obtaining answers at a very large number of points on the surface is no longer a problem. Therefore sufficiently detailed results can be obtained to permit their presentation in the form of maps of the surface, where the x- and y-directions represent distances on the surface, and the z-direction adsorption potential contours. Thus the difficulty now lies in the formulation of an adequate physical and mathematical model on which to base the calculations, and all the difficulties that this presents have by no means been resolved. To illustrate the methods and the problems, even with a comparatively simple case, we will look at the results obtained by Ricca *et al* [88-90] for the adsorption of an inert gas molecule on a solid inert gas substrate.

The calculations were based on the assumption that the total interaction energy can be evaluated by adding the pair-wise values of the interaction between the adsorbing atom and each atom in the substrate. The individual values were calculated using the Lennard-Jones '6-12' potential equation

$$U(r) = \epsilon_0 \left[\frac{r_0}{r} \right]^{12} - 2 \left[\frac{r_0}{r} \right]^6 \quad , \tag{4.42}$$

where ϵ_0 is the depth of the energy minimum and r_0 is the value of the separation, r, at this point. An example of their results is shown in Fig. 4.26 as a contour map of the potential energy surface, showing that the adsorption site is at the mid-cell position[†], the minimum translational barrier being over the saddle

[†] The calculations assumed ideal bulk crystal positions throughout, that is, no relaxation due to the presence of the surface, and no secondary relaxation as adsorption occurs.

point to a neighbouring site. In order to estimate the degree of localisation, or state of mobility, of the adsorbed atoms of the surface, Ricca stated that simple comparison of mean translational energy levels with translational barrier heights was inadequate (cf Hill's hindered translation model – section 4.6.8). The exact nature of the surface can be better visualised by the section illustrated in Fig. 4.27 which represents the vertical plane through Fig. 4.26 at the position represented by a horizontal line joining A-A and passing through S, halfway up the contour map. It should be noted that the equilibrium adsorption distance from the surface plane is a positional variable.

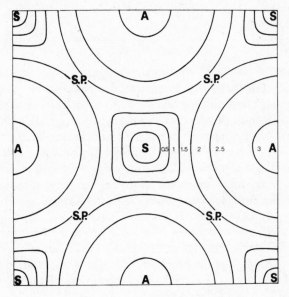

Fig. 4.26 – Map of the adsorption potential energy surface for an He atom on the (100) face of solid Xe. Numbers on the energy contours give energies related to the ϵ_0 value for an isolated He-Xe pair of atoms.
A = surface atom,   S = adsorption site   S.P. = saddle point.
(Reproduced with permission from Ricca *et al* [88]).

Ricca *et al* suggested that a better way to approach the concept of mobility, would be to use a quantum mechanical approach to calculate probability densities and vibrational energy levels for adsorbed gases by solving the Schrödinger equation for an atom in the already determined gas-solid potential field:

$$H\psi = -\frac{h^2}{2m}\nabla^2\psi + {}_mU_o(\bar{r})\psi = \epsilon\psi \tag{4.43}$$

where H is the Hamilton operator and ϵ and ψ are, respectively, the energy eigenvalue and eigenfunction of the adsorbed atom. On an ideal inert-gas crystal surface, $_mU_o(\overline{r})$ is a periodic function in two dimensions, and consequently $\psi(\overline{r})$ and the probability density $\psi\psi^*$ (where ψ^* is the complex conjugate of ψ), are also periodic. Thus calculating the local wave function for an adsorbed over a single site would be a useful first step, since the combination of such functions could give the wave function for the entire surface.

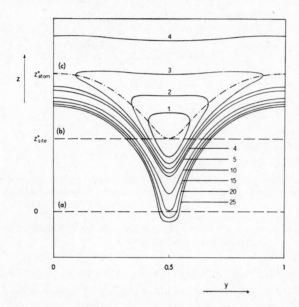

Fig. 4.27 – Section through the potential energy surface illustrated in Fig. 4.26. The He atom has variable y,z coordinates in a plane of constant x that is the bisector of the unit cell. a is the cell parameter. Iso-potential energy curves labelled as in Fig. 4.26. (a) represents the surface plane, (b) represents the plane containing all surface sites and the curve (c) indicates how the locus of the energy minimum moves from point A to point S on Fig. 4.26. (Reproduced with permission from Ricca *et al* [88]).

The results of such calculations indicate that the energy levels for an atom on a surface containing N_s sites are concentrated into narrow bands containing N_s levels each. The width of a band depends upon the mass of the adsorbed atom as well as the parameters involved in determining $_mU_o$. Steel [91] has given an energy band scheme based on Ricca's results, and this is shown in Fig. 4.28. This type of diagram would be valid for any adsorbed atom, and adsorbed helium is only a special case because the bandwidths of the low-lying states are much larger than for heavier atoms.

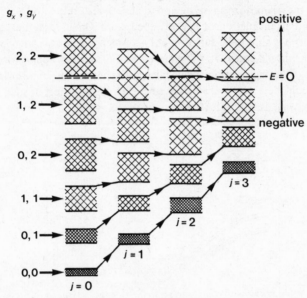

Fig. 4.28 – Schematic diagram of selected energy *bands* for an adsorbed atom.
Some of the translational bands for several values of the quantum number j for
motion perpendicular to the surface are shown, with the density of levels within
a band roughly indicated by the density of the cross-hatching. The zero energy
is defined as the ground state energy of a free atom. Bands with identical g values
are connected by arrows. (After Steele [91]).

The results in terms of probability density diagrams are shown in Fig. 4.29
for various inert gas atoms on the (111) face of a xenon crystal. These represent
the fundamental 'bonding' state of two adjacent sites. A qualitative difference
exists between the localised atoms of Ar and Ne and the mobile atoms of the
He^4 and He^3. The 'orbitals' for the He atoms extend over both sites in contrast
to the Ar and Ne cases.

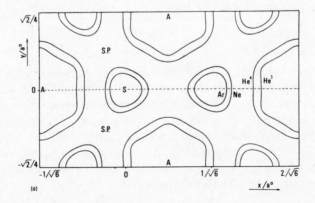

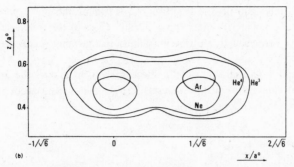

Fig. 4.29 – Isodensity surfaces including 95% of the total probability for the fundamental 'bonding' state of various inert-gas atoms adsorbed on the (111) face of a Xe crystal. (a) Horizontal section through the average z-ordinate for each gas. (b) Vertical sections on a plane joining two adjacent adsorption sites. Symbols as in Fig. 4.26. a_o is the cell parameter. (Reproduced with permission from Ricca *et al* [90]).

4.6.4.2 *Adsorption potential maps; ionic solids*

The calculation of adsorption potentials on an ionic solid presents considerable problems. The exact form of the equation used to represent the dispersion force interactions between the adsorbing atom or molecule is somewhat difficult to decide, and there is the added contribution to the total interaction of the electric field due to the charges on the ions. This latter contribution is not an insignificant proportion of the total, and if realistic assessments are to be made, the complication of surface relaxation cannot be ignored. The data most suited to illustrating these points are those of House and Jaycock [92,93] for the adsorption of inert gas atoms on ideal and relaxed alkali halide crystals.

The total interaction between an ionic crystal and an adsorbing atom can be written as

$$_m U_o = U^d(r) + U^e \qquad (4.44)$$

where $U^d(r)$ is the dispersion-repulsion interaction energy and can be estimated by the summation of the pairwise interactions over all the ions in the solid,

$$U^d(r) = \Sigma U(r)$$
$$\text{all ions}$$

although it was found that summation over the nearest 2000 ions was sufficiently accurate. $U(r)$ was given by the '6-exp' equation in the Buckingham form

$$\frac{U(r)}{k} = \frac{\epsilon/k}{1-(6/\alpha')} \left\{ \frac{6}{\alpha'} \exp\left[\alpha'(1-(r/r_m))\right] - (r/r_m)^6 \right\} \text{ for } r > r_{max} \qquad (4.45)$$

$$\frac{U(r)}{k} = \infty \text{ for } r < r_{max} \qquad (4.46)$$

where ϵ is the minimum in the function, r_m is the interatomic separation at the minimum, α' is a constant determining the steepness of the repulsive part of the curve, and r_{max} is the position of the maximum in the function occurring at small separations.

If α is the polarisability of the inert gas atom, and F the electric field modulus at the point above the crystal surface where the atom is, then the electrostatic potential energy is given by

$$U^e = - \int_0^F \alpha F \, dF = - \alpha F^2 / 2 \qquad (4.47)$$

since α is isotropic for spherically symmetrical atoms. In principle, it would be possible to determine the electric field vector by summation of the individual ion contributions, but this method is so slowly convergent involving many tens of thousands of terms, that the method first proposed by Madelung [94], adopted by Born [95] and Lennard-Jones and Dent [96], was slightly modified by House and Jaycock [92].

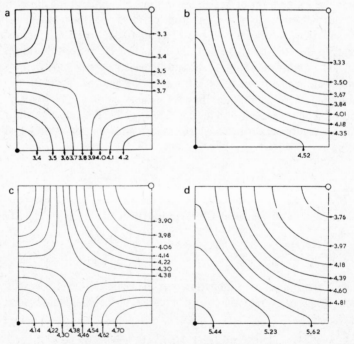

Fig. 4.30 — Isopotential energy curves, values shown in units of kJ mol^{-1}, for the interaction of (a) Ar on the unrelaxed (100) face of NaCl, (b) Ar on the relaxed (100) face of NaCl, (c) Kr on the unrelaxed (100) face of NaCl, and (d) Kr on the relaxed (100) face of NaCl. ● Na$^+$, ○ Cl$^-$. (Reproduced with permission from House and Jaycock [93]).

The method deals adequately with an ideal crystal geometry, but cannot cope with the perturbation of crystal structure near the surface due to relaxation. House and Jaycock [93] used the method to account for the contribution to the electric field due to the bulk of the crystal, but used separate Fourier type expansions of each of the outermost five layers, based on the geometric calculations of Benson *et al* [16].

The results of the calculations are illustrated in the form of adsorption potential maps in Figs. 4.30 and 4.31. They illustrate one most important fact, namely that surface relaxation can completely alter the map from that obtained for an ideal surface. Not only are the values of the adsorption potentials changed, quite appreciably in certain cases, but also the position of a maximum in $_m U_o$ can also be different. This latter point is clearly seen in Fig. 4.30, where the maximum value of $_m U_o$ for Ar on the unrelaxed (100) face of NaCl is at the midcell position, and on the corresponding relaxed face is over the Na^+ ion.

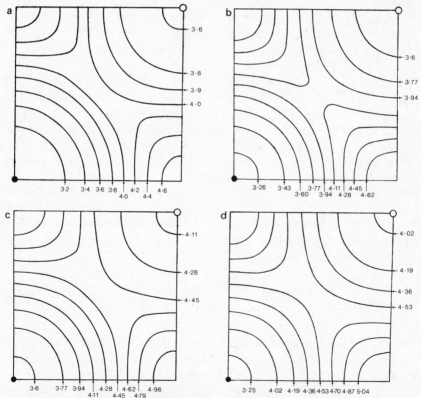

Fig. 4.31 – Isopotential energy curves, values shown in units of kJ mol^{-1}, for the interaction of (a) Ar on the unrelaxed (100) face of KCl, (b) Ar on the relaxed (100) face of KCl, (c) Kr on the unrelaxed (100) face of KCl and (d) Kr on the relaxed (100) face of KCl. ● K^+, ○ Cl^-. (Reproduced with permission from House and Jaycock [93]).

Whilst the effects of surface relaxation are immediately obvious in the case of NaCl, the situation is not so clear for KCl. The relaxed and unrelaxed cases shown in Fig. 4.31 show very similar values for both Ar and Kr adsorbed on the (100) face of KCl. The results are summarised in Table 4.5.

Table 4.5

Comparison of the maximum value of $_mU_o$ and the barrier to translation, V_o, on relaxed and unrelaxed (100) faces of NaCl and KCl [93].

System	$_mU_o$ (max)/kJ mol^{-1}		position of $_mU_o$		V_o/kJ mol^{-1}	
	unrelaxed	relaxed	unrelaxed	relaxed	unrelaxed	relaxed
Ar/NaCl	4.284	4.641	mid-cell	over Na$^+$	0.519	0.080
Kr/NaCl	4.785	5.449	mid-cell	over Na$^+$	0.450	0.457
Ar/KCl	4.703	4.818	mid-cell	mid-cell	0.695	0.722
Kr/KCl	5.074	5.197	mid-cell	mid-cell	0.623	0.646

Some idea of what these maps indicate in terms of the mobility of an adsorbed film can be gained using Hill's hindered translation model [97]. An estimate of the state of an adsorbed atom may be obtained by consideration of quantum energy states of a one-dimensional oscillator located at the potential energy minimum. The fraction of oscillators [98] with energy E^{vib} greater than or equal to $(n + \frac{1}{2})hv^{\perp}$ is given by $\exp(-nhv^{\perp}/kT)$. The smallest value of the quantum number n satisfying the condition

$$(n + \tfrac{1}{2})hv^{\perp} \geqslant E^{vib} \tag{4.48}$$

may be used to determine the fraction R_t of adsorbed atoms translating $(_mU_o > E^{vib} > V_o)$ which is given by

$$R_t = \frac{\exp(-n_1 hv^{\perp}/kT) - \exp(-n_2 hv^{\perp}/kT)}{1 - \exp(-n_2 hv^{\perp}/kT)} \tag{4.49}$$

where n_1 and n_2 are given by equation (4.48) with E^{vib} equal to V_o and $_mU_o$ respectively. The classical equation analogous to (4.49) is a poor approximation in this case.

The z-component of the vibrational frequency, $v^{\perp}$, may be determined directly from the 'best fit' parabola for the total potential energy U around $_mU_o$ and making the simple harmonic approximation

$$f_z = (\partial^2 U/\partial z^2)_{x,y,z^o} \tag{4.50}$$

giving

$$v^{\perp} = (1/2\pi)(f_z/m)^{\frac{1}{2}} \qquad\qquad (4.51)$$

where f_z is the force constant and m the mass of the adsorbed atom. Similar expressions are available for v_x and v_y, the x and y direction vibrational frequencies. The results of such calculations for the data shown in Figs. 4.30 and 4.31 are shown in Table 4.6, and they indicate that a significant number of the molecules would be unable to translate in three out of the four cases. Unfortunately, since these results are dependent upon the accuracy of the original adsorption potential calculations, their validity is not known at the present time. The model can also be used to predict isosteric heats, and yields 4.48 kJ mol^{-1} for Ar on NaCl, and 4.55 kJ mol^{-1} for Ar on KCl. The best experimental values available at present are those of Hayakawa [99], whose corresponding extrapolated values are 9.17 and 8.71 kJ mol^{-1}. Although this agreement is poor, it should be remembered that surface heterogeneity will tend to increase experimental values at low coverages. Nevertheless these calculations are limited primarily by the uncertainty in the form of, and constants for, the inter-atomic potential, as well as the validity of the pair-wise summation procedure adopted which must of necessity ignore multibody interactions. It still must be said that they represent the only attempt so far to investigate adsorption on relaxed ionic crystals, remembering that they also ignore secondary relaxation on adsorption which is likely to be a small effect in this case.

Table 4.6

Values of R_t, the ratio of translating to non-translating adsorbed atoms calculated from the hindered translation model.

System	$v^{\perp} \times 10^{12}/\text{s}^{-1}$	$_mU_o/\text{kJ mol}^{-1}$	$V_o/\text{kJ mol}^{-1}$	R_t
Ar/NaCl	1.108	4.461	0.080	1.00
Kr/NaCl	0.750	5.449	0.457	0.40
Ar/KCl	1.004	4.818	0.722	0.29
Kr/KCl	0.681	5.197	0.646	0.43

Unfortunately the alkali halides are the only compounds for which detailed calculations for the relaxation behaviour are available. Thus all other calculations are hedged about with even more uncertainties. One or two examples are nevertheless interesting, for example the adsorption of water on rutile [74]. The adsorption of a water molecule on an ionic solid is complicated by the fact that the molecule has a dipole and is triatomic. Therefore the calculation of $_mU_o$ must allow the molecule to rotate to its minimum energy position, if the results are to be of any significance. The results for the (110) face are shown in Fig. 4.32.

These results are obviously open to many objections as previously suggested, but the fact that they indicate very high values of $_mU_o$, of the order of 430 kJ mol^{-1} on the (110) face and 410 kJ mol^{-1} for the (100) face is interesting. Perhaps giving an insight into why the chemisorptive change to two surface hydroxyl groups appears to occur quite readily.

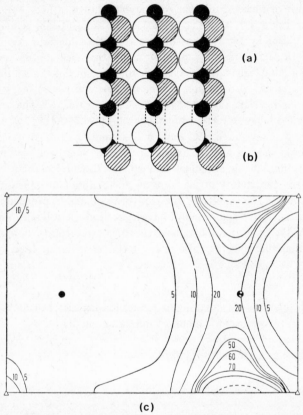

Fig. 4.32 – Adsorption of water on the (100) face of rutile. (a) A plan view of the (100) cleavage plane. There are 7.4 Ti ions nm^{-2}. (b) An elevation looking along the c-axis in the line of the (100) cleavage plane. ◐, ○, O; ●, Ti. (c) $_mU_o$ contour map, the contours in kcal mol^{-1}, △, Ti^{4+} ion in surface; ●, O^{2-} ion above surface; ◓, O^{2-} ion below surface. (Reproduced with permission from Jaycock and Waldsax [74]).

4.6.4.3 *Mobile, localised and pseudo-localised adsorption*

One of the more frequently quoted visualisations of these terms is due to Ross [100] and is illustrated in Fig. 4.33. In his diagram (a) represents mobile adsorption, where V_o is small and also much less than $_mU_o$, (b) represents localised adsorption where $_mU_o = V_o$ and it is only possible for a molecule to migrate from site to site through the bulk gas phase, and (c) is the pseudo-localised case, where the translational energy band width, $_mU_o - V_o$, is small, resulting in few adsorbed molecules being able to translate. Although this diagram is convenient, it is in certain respects unrealistic, and there are alternative methods of looking at the situation which give more insight.

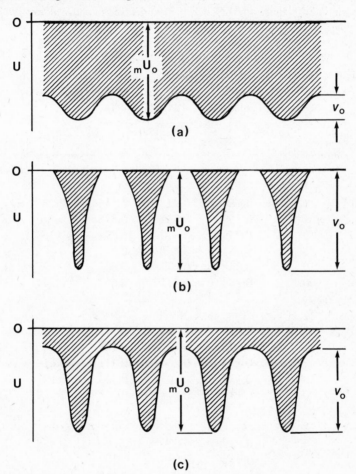

Fig. 4.33 – Ross's illustration [100] of (a) mobile, (b) localised and (c) pseudo-localised models of physical adsorption.

Let us return to equation (4.49) which gives an expression for the fraction of molecules translating. Thus the truly mobile case will correspond to $R_t = 1$, the localised case to $R_t = 0$, and all intermediate values of R_t to varying degrees of pseudo-localisation. Thus mobile adsorption implies that $n_1 = 0$ and localised adsorption that $n_1 = n_2$. The former case can be found in Table 4.6 for the system Ar/NaCl, but the latter case has not been found in any of these cases for which detailed calculations are available, and seems a somewhat unlikely occurrence. To illustrate the point further, Table 4.7 shows the variation of the fraction translating with the vibrational quantum numbers n_1 and n_2, over a range of values of $h\upsilon^{\perp}/kT$, corresponding approximately to the range of values that were found by House and Jaycock [93] and Jaycock and Waldsax [74].

Table 4.7

Calculated values of the fraction of adsorbed atoms translating, R_t, for various values of $h\upsilon^{\perp}/kT$, and selected values of the quantum numbers n_1 and n_2. (Equations (4.48) and (4.49)).

n_1	n_2	$h\upsilon^{\perp}/kT$ 1.00 R_t	0.75 R_t	0.50 R_t	0.30 R_t	0.10 R_t
1	1	0	0	0	0	0
	1.5	0.186	0.219	0.254	0.285	0.317
	2	0.269	0.321	0.378	0.426	0.475
	3	0.335	0.410	0.494	0.563	0.633
	5	0.364	0.460	0.571	0.666	0.758
	10	0.368	0.472	0.604	0.727	0.849
	100	0.368	0.472	0.607	0.741	0.905
2	2	0	0	0	0	0
	2.5	0.058	0.082	0.114	0.145	0.181
	3	0.090	0.132	0.186	0.240	0.301
	5	0.129	0.204	0.311	0.419	0.539
	10	0.135	0.223	0.364	0.525	0.713
	100	0.135	0.223	0.368	0.549	0.819
3	3	0	0	0	0	0
	3.5	0.020	0.036	0.060	0.087	0.122
	4	0.032	0.059	0.102	0.151	0.214
	5	0.043	0.084	0.154	0.236	0.341
	10	0.050	0.105	0.218	0.375	0.590
	100	0.050	0.105	0.223	0.407	0.741
5	5	0	0	0	0	0
	7.5	0.006	0.020	0.060	0.132	0.254
	10	0.007	0.023	0.076	0.182	0.378
	100	0.007	0.024	0.082	0.223	0.607

This table illustrates four important points:

(i) As n_1 increases the degree of localisation increases, that is, increasing translational barriers increases the degree of localisation;

(ii) For a fixed value of n_1, increasing values of n_2 increases the mobility of the adsorbed film, that is, increasing the width of the translational energy band increases mobility;

(iii) Decreasing the size of the vibrational quanta increases mobility;

(iv) Increasing temperature increases the film mobility.

On the basis of such calculations as are currently available, many cases would be expected to be pseudo-localised, and a few to correspond to the mobile and localised extremes. Of the cases so far discussed, H_2O on the surface shown in Fig. 4.32 would be localised (calculations suggest $R_t \sim 10^{-50}$), and Ar/NaCl mobile, but of the other cases illustrated in Figs. 4.30 and 4.31, and Table 4.6, namely Kr/NaCl, Ar/KCl and Kr/KCl, all are pseudo-localised. It is interesting to note that increasing the size of the adsorbed atom does not necessarily increase the degree of localisation. For example, Ar/NaCl is mobile and 40% of the adsorbed atoms translate for Kr/NaCl; but 29% of adsorbed Ar atoms on KCl translate, compared with 43% in the case of Kr/KCl (see Table 4.6).

4.6.5 Monolayer Models

Before proceeding to the discussion of the relationship between experimental adsorption isotherms and physical models of the adsorption process, it is essential to consider the ways in which adsorption data are presented. Probably because of the way in which early workers in the field, such as Langmuir, represented their data, there has evolved the use of terminology which is proving most difficult to supersede. The simplest way to represent the amount of a particular molecule adsorbed at a particular temperature and equilibrium pressure would be as either the number of moles, n^s, or molecules, N^s, in the adsorbed layer, thus

$$N^s/L = n^s = f(P,T) \tag{4.52}$$

but it is often expressed instead as the equivalent volume of n^s moles of adsorbate at STP, V^s

$$V^s = 22\,414\,n^s = f(P,T) \tag{4.53}$$

where V^s is now in cm^3 at STP. Therefore an adsorption isotherm may be represented as either

$$V^s_T = f(P) \tag{4.54}$$

or

$$n_T^s = f(P) \quad .$$ (4.55)

It is possible to describe a monolayer in an exactly analogous way to that used for an insoluble monolayer at the air-water interface (section 3.11.4), and consider what equations of state might be adopted for the adsorbed film. The ideal 2-d gas equation would suggest itself as a starting point,

$$\Pi A = kT$$ (3.148)

and the problem becomes one of relating the parameters which cannot be measured directly, Π and A, to those which can, P and n^s.

We may write the Gibbs adsorption isotherm (2.48) for the adsorption of a one component gas as

$$-d\gamma = \Gamma \, d\mu$$ (4.56)

which on substitution for Γ yields

$$-d\gamma = \frac{n^s}{A} \, d\mu = \frac{n_m^s}{A} \cdot \frac{n^s}{n_m^s} \, d\mu \quad .$$ (4.57)

However, the quantity n^s/n_m^s, where n_m^s is the amount in a complete monolayer, is the fractional coverage, θ, thus

$$-d\gamma = \frac{n_m^s}{A} \theta \, d\mu \quad .$$ (4.58)

The chemical potentials in the adsorbed and equilibrium gas phase must be equal, therefore

$$\mu = \mu_{gas} = \mu^{\ominus} + kT \ln P$$ (4.59)

and

$$d\mu = kT \, d(\ln P) \quad .$$ (4.60)

Substitution into equation (4.58) yields

$$-d\gamma = \frac{n_m^s}{A} \, kT \, \theta \, d(\ln P) \quad .$$ (4.61)

If this equation is integrated with the boundary conditions that $\gamma = \gamma_0$ when $P = 0$, and the upper limit that the surface tension is γ when the pressure is P, then we may write

$$\int_{\gamma_0}^{\gamma} -d\gamma = \frac{n_m^s}{A} kT \int_0^P \theta \, d(\ln P) \quad . \tag{4.62}$$

Then by writing $A_m = A/n_m^s$, and using the definition of the surface pressure, $\Pi = \gamma_0 - \gamma$, we have

$$\Pi = \frac{kT}{A_m} \int_0^P \theta \, d(\ln P) \quad . \tag{4.63}$$

The ideal 2-d gas equation (3.148) may be rewritten

$$\Pi = \frac{kT}{A} = \frac{kT}{A_m} \cdot \frac{A_m}{A} = \frac{kT}{A_m} \theta \tag{4.64}$$

Therefore differentiating and using equation (4.63) we have

$$d\Pi = \frac{kT}{A_m} \, d\theta = \frac{kT}{A_m} \theta d(\ln P) \tag{4.65}$$

or

$$d(\ln \theta) = d(\ln P)$$

whence $P = K\theta$ \hfill (4.66)

which is known as Henry's Law, where K is a constant.

The derivation of Henry's Law just given is a particular example of the manipulation of a 2-dimensional equation of state, which may in general be written

$$\Pi = \frac{kT}{A_m} f(\theta, T) \quad . \tag{4.67}$$

This would give

$$\ln P = \int_0^\theta f'(\theta, T)_T \, d(\ln \theta) + \ln K \tag{4.68}$$

where K is an integration constant, and the generalised model isotherm function would be

$$P = K g(\theta, T) \quad . \tag{4.69}$$

This has been well presented by Ross and Olivier [101].

The first improvement upon the ideal equation is to make allowance for the finite size of the adsorbing species, as in equation (3.129),

$$\Pi(A - A_0) = kT \tag{3.129}$$

which using the above approach yields the Volmer adsorption isotherm,

$$P = K \frac{\theta}{1 - \theta} \exp\left(\frac{\theta}{1 - \theta}\right) \quad . \tag{4.70}$$

The further extension of the equation of state to the 2-d analogue of the van der Waals equation

$$\left[\Pi + \frac{\alpha}{A^2}\right](A - \beta) = kT \tag{4.71}$$

where α and β are constants, gives the adsorption isotherm

$$P = K \frac{\theta}{1 - \theta} \exp\left[\frac{\theta}{1 - \theta} - \frac{2\alpha\theta}{kT\beta}\right] \tag{4.72}$$

often known as the Hill-de Boer equation [102].

The constant K in equations (4.66), (4.70) and (4.72), may be related to the adsorption potential ${}_mU_0$ by

$$K = A^0 \exp\left(-{}_mU_0/RT\right) \tag{4.73}$$

where

$$\ln A^0 = -\frac{\Delta S^a}{R} + \frac{E^{vib} - E_0^{vib}}{RT} + \frac{\Delta E^{tr}}{RT} + \frac{\Delta E^{rot}}{RT} + \frac{\Delta E^{ia}}{RT}$$

$$- \ln g(\theta, T) + \ln P^{\ominus} - 1 + \frac{A_m}{R\theta}\left[\frac{\partial \Pi}{\partial T}\right]_\theta \quad . \tag{4.74}$$

In this equation ΔS^a is the integral entropy of adsorption, ΔE^{tr} and ΔE^{rot} are the translational and rotational energy changes on adsorption, ΔE^{ia} is the potential energy of interaction of the adsorbate, and $P^{\ominus}$ is the standard pressure. Therefore A^0 can be evaluated for a particular isotherm function, $g(\theta, T)$.

These three theoretical isotherms are based on essentially the same assumptions as those used in the kinetic energy theory of 3-d gases, and therefore are descriptions of mobile 2-d adsorbed films. The Henry's Law case would only be expected to apply at very low coverages, and as the coverage increases Volmer's equation would apply. However, neither of these two equations predict phase transitions which can be observed in certain cases, as for example the adsorption of krypton on sodium chloride [103] shown in Fig. 4.34. The Hill-de Boer equation, with appropriate values of the constants α and β, does predict such behaviour. The critical point corresponds to $2\alpha/kT\beta = 6.75$, when $\theta = 1/3$ and $P/K = 0.08689$. Above the critical value of $2\alpha/kT\beta$ the equation has one real and two imaginary roots and therefore predicts a unique value for the amount adsorbed at a particular pressure, but below the critical value of $2\alpha/kT\beta$ there is the possibility of there being 3 real roots, as can be seen in Fig. 4.35. For example the isotherm for $2\alpha/kT\beta = 10$, in fact would follow the solid line, whereas the equation predicts the dashed curve. The phase transition between the point g, where the film is a 2-d gas, and the point 1, where it is a 2-d liquid, occurs at the pressure where the chemical potentials of the gas and liquid phases are equal [104].

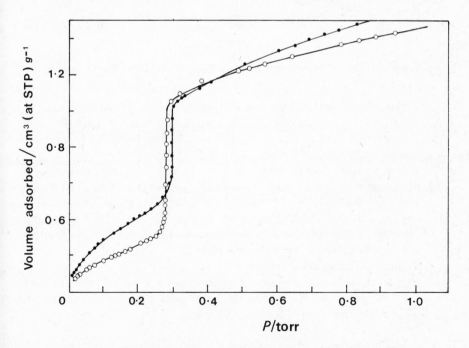

Fig. 4.34 – Adsorption isotherms at 76.1 K for krypton on sodium chloride.
(From House and Jaycock [103]).

•, experimental points obtained before annealing,
○, experimental points obtained after 21 hours' annealing.

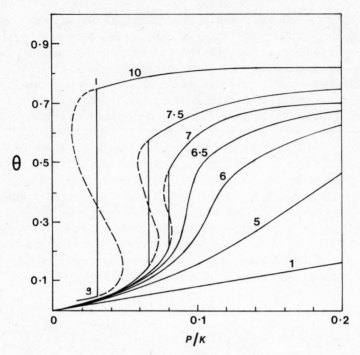

Fig. 4.35 – The Hill-de Boer equation (4.72) plotted for various values of $2\alpha/kT\beta$.

Models for localised monolayers cannot readily be derived from considerations of the corresponding equations of state, although the latter do exist. The first approach was by Langmuir [105], who used a kinetic derivation. The assumptions upon which it is based have been clearly stated by Fowler [106].

"necessary and sufficient assumptions. . . . are that the atoms (or molecules) of the gas are adsorbed as wholes onto definite points of attachment on the surface of the adsorber, that each point of attachment can accommodate one and only one adsorbed atom, and that the energies of the states of any adsorbed atom are independent of the presence or absence of other adsorbed atoms on neighbouring points of attachment".

This defines an ideal localised monolayer.

Langmuir considered the dynamic equilibrium between the adsorbed layer and the bulk gas phase. The kinetic theory of gases gives that the number of molecules striking unit area of surface in unit time as

$$i = P/(2\pi mkT)^{\frac{1}{2}} \quad . \tag{4.75}$$

Adsorption will not occur if a molecule hits a site that is already occupied, and of those molecules hitting a bare site, only a certain fraction, j, will be captured (that is, adsorbed), the remainder being reflected. If the fraction of sites already occupied is θ, then the rate of adsorption will be $ij(1 - \theta)$. The rate at which molecules return to the gas phase from the adsorbed film can be written as $l\theta$, where l is the rate of evaporation from a fully covered surface. At equilibrium the number of molecules in the adsorbed film must be a constant, therefore

$$ij(1 - \theta) = l\theta \quad . \tag{4.76}$$

Substituting for i in equation (4.76) we have

$$\theta = \frac{bP}{1 + bP} \tag{4.77}$$

or

$$bP = \frac{\theta}{1 - \theta} \tag{4.78}$$

where

$$b = \frac{j}{l(2\pi mkT)^{\frac{1}{2}}} \tag{4.79}$$

which are representations of the Langmuir adsorption isotherm.

This isotherm has been deduced many times and in many different ways [107], but the thermodynamic approach [108,109] leads to an explicit definition of the constant b:

$$b = \frac{h^3}{kT \, (2\pi mkT)^{3/2}} \frac{f_a(T)}{f_g(T)} \, \exp \, (q/kT) \tag{4.80}$$

where $f_a(T)$ and $f_g(T)$ are the internal partition functions for a molecule in the adsorbed and gaseous state respectively, and q is the energy required to transfer a molecule from the lowest adsorbed state to the lowest gaseous state. It should be noted that it is possible to transpose the Langmuir isotherm into the form of equation (4.69) where $K = 1/b$ and $g(\theta,T) = \theta/(1 - \theta)$.

As was stated earlier in Fowler's description of an ideal localised monolayer, the Langmuir isotherm specifically excludes lateral interaction between adsorbed molecules on neighbouring sites. Obviously this interaction must exist, and a statistical thermodynamic approach [110] leads to

$$P = K \frac{\theta}{1 - \theta} \, \exp - \left[\frac{e\theta}{kT} \right] \tag{4.81}$$

where ϵ is the lateral energy of interaction [111]. The usual derivation replaces ϵ with $c\omega/2$, where c is an assumed co-ordination number of the underlying lattice and ω is the pair interaction energy for molecules on neighbouring sites. Equation (4.81) is usually referred to as the Fowler-Guggenheim equation. This form of equation produces isotherms very similar in form to the Hill-de Boer equation, as can be seen in Fig. 4.36. In the Fowler-Guggenheim case the critical condition corresponds to $-\epsilon/kT = 4$, $\theta = 0.50$ and $P/K = 0.1353$.

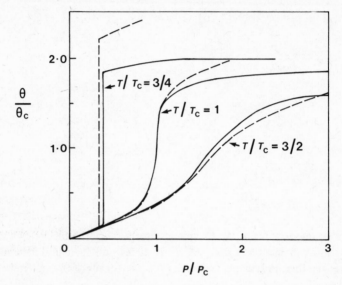

Fig. 4.36 – Comparison of theoretical isotherms in reduced units for the Hill-de Boer equation (4.72) and the Fowler-Guggenheim equation (4.81). θ_c, T_c and P_c are the values at the theoretical critical points for 2-d condensation. The values of these parameters are not the same for each model.
–––– Hill-de Boer, ——— Fowler-Guggenheim.

The prime limitation of all these models in relation to their comparison with experimentally determined isotherms is that they assume a single value for K, that is the surface is homogeneous or at the very least homotattic. In any real case the surface is heterogeneous, although the overwhelming majority of workers choose to overlook the fact.

4.6.6 Determination of surface heterogeneity
In the methods about to be described, an attempt is made to use the monolayer models discussed in the previous section and an experimentally determined isotherm, to deduce a distribution of adsorption potentials. Such a distribution will be in part dependent on the nature of the substrate surface, but there will also be an effect due to the nature of the adsorbate gas. There are in essence two approaches in the literature:

(i) To calculate isotherms from postulated distributions and then compare these with the experimental isotherm. In practice this is limited by the scope of the prior determined comparison isotherms [112].

(ii) To make use of the calculating power of modern computers to speed up an approach somewhat similar to that of Adamson [113], in which a distribution is deduced by a successive approximation technique from the experimental isotherm. This has been done by Ross and Morrison [114] and by House and Jaycock [115]. The method of the latter authors will be used as an example.

The algorithm HILDA (Heterogeneity Investigated by a Loughborough Distribution Analysis) was developed to determine the adsorption energy distribution function from the integral equation for the adsorption isotherm on a 'patchwise' heterogeneous surface:

$$V^s = V^s_m \int_0^\infty F'(U)\theta(P,U)dU \qquad (4.82)$$

where V^s is the amount of gas adsorbed at the equilibrium pressure, P; $F(U)$ is a numerical function of the fraction of the surface with energies U or less, hence $dF(U)/dU = F'(U)$, is the frequency of 'patches' per unit energy interval; $\theta(P,U)$ is the fractional coverage of a unisorptic patch given by the model isotherm function; U is equal the the maximum adsorption energy, $_mU_o$, minus the zero state vibrational energy, and is taken as positive in magnitude; V^s_m may be considered in this case as a normalising factor such that

$$\int_0^\infty F'(U)dU = 1 \quad . \qquad (4.83)$$

A direct application of equation (4.82) is not practicable since some evaluation of the limits of U must be obtained. The method adopted in HILDA was originally suggested by Berenyi [116]. If the temperature and adsorbate-adsorbent interactions are such that the Hill-de Boer or Fowler-Guggenheim model isotherm is subcritical, a phase transition will be located at (P/K), and hence given the lowest pressure P_l of the data set P, the corresponding K_l value (related to the adsorption energy) may be obtained on the energy 'patch' where the adsorbate will be at the phase transition. This K_l is then used to determine the upper limit to the adsorption energy distribution that will be computed, U_h. 'Patches' of energy $> U_h$ are assumed to be full. This approximation avoids the difficulty of proposing a distribution of adsorption energies for U in the range U_h to ∞. Another method would be to determine the amount of adsorption at a particular pressure through equation (4.82) by assuming that there are no 'patches' of energy $> U_h$. As expected, this leads to an increase in F' at U_h in the final distribution. Obviously information concerning F' is lost near U_h if the isotherm data are not extensive enough; in general, a $V_i^s/V_N^s < 0.05$, where N is the total number of data points, is found to be satisfactory. The lower limit, U_l may be obtained from the highest pressure, P_N. 'Patches' of energy $< U_l$ are assumed to be empty. If the model isotherm is supercritical the U_h and U_l limits are obtained by determining $(P/K)_t$ from the point of inflection of the Hill-de Boer and Fowler-Guggenheim model isotherm functions or the value at $\theta = 0.5$ for the Langmuir isotherm. The values of $(P/K)_t$ are available for the Fowler-Guggenheim and Hill-de Boer functions [115,117].

Using an initial approximation to $F(U)$, the problem is to reduce the root mean square (rms) deviation from the experimental data to a minimum.

$$\text{rms} = N^{-1} \left\{ \sum_{i=1}^{N} \left[\frac{(V_i^s/V_m^s) - \int_{\theta_l(P_i)}^{\theta_h(P_i)} F(U)\mathrm{d}\theta + F_1 \left[1 - \theta_h(P_i)\right] + F_N\theta_l(P_i)}{V_i^s/V_m^s} \right]^2 \right\}^{1/2}$$

(4.84)

The iterative improvement is obtained by adjusting $F(U)$:

$$F_i^c = F_i^{c-1} \left\{ \frac{V_i^s/V_m^s}{\int_{\theta_l(P_i)}^{\theta_h(P_i)} F^{c-1}(U)\mathrm{d}\theta + F_1^{c-1} \left[1 - \theta_h(P_i)\right] + F_N^{c-1} \theta_l(P_i)} \right\}$$

(4.85)

with the condition that if

$$F_i^c < F_{i-1}^c$$

(4.86)

then

$$F_i^c = F_{i-1}^c + 10^{-8}$$

(4.87)

and normalising:

$$V_m^{s\,c} = V_m^{s\,c-1} \times F_N$$

(4.88)

$$F_i^c = F_i^{c-1}/F_N$$

(4.89)

where the superscript c labels the cycle. The values of θ_h and θ_l are determined from the isotherm function using the limiting values of P. In this method all the F_i^c, values are changed after each iteration to obtain a new set of F_i^{c+1} values, with the V_m^s from equations (4.88) and (4.89) until the rms reaches a minimum.

For a rapid convergence using this procedure the choice of the initial $F(U)$ is important. The method of Adamson and Ling [113] chooses the amount adsorbed at monolayer coverage from a BET (see sections 4.6.7 and 4.6.8) analysis, $V_m^s (\text{BET})$, and determines the first approximation to $F(U)$ through

$$F_i^1 = V_i^s / V_m^s (\text{BET}) \quad . \tag{4.90}$$

This method suffers from the disadvantage that $V_m (\text{BET})$ is not always known or reliable and that the subsequent distribution function, F', is not normalised, that is, $F_n \neq 1$. The normalising procedure (equations (4.88) and (4.89)) used here allows an arbitrary choice of V_m^s which is adjusted in subsequent cycles to give $F_N = 1$. The initial approximation adopted here is

$$F_i^1 = V_i^s / V_N^s \tag{4.91}$$

$$V_m^{s\,1} = V_N^s \quad . \tag{4.92}$$

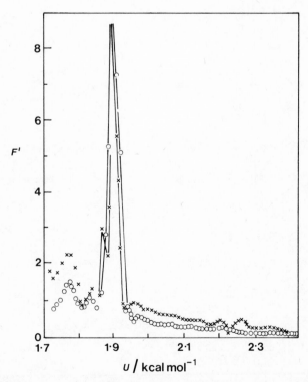

Fig. 4.37 – Site energy distribution function obtained by HILDA [118] from the analysis of the data shown in Fig. 4.34 using the Hill-de Boer isotherm function. X unannealed sample. ○, annealed sample.

The algorithm used in HILDA represents the $F(U)$ and $F(K)$ points by cubic splines. Facilities are available for smoothing the data and making multilayer adsorption corrections. The integrals shown in equations (4.84) and (4.85) are evaluated using the numerical method of Clenshaw and Curtis. The final distribution function is determined in the range of U_l to U_h. The function F' is generated either by using the cubic spline fits to $F(U)$ or by determining the gradient by a linear interpolation between consecutive points.

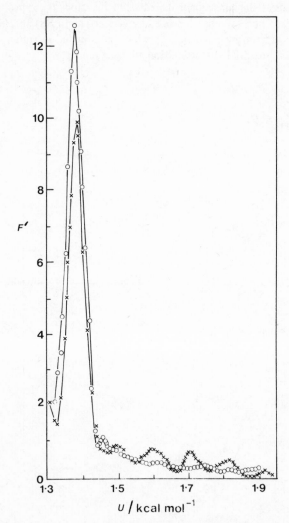

Fig. 4.38 – Site energy distribution obtained by HILDA [118] from the analysis of argon adsorption data on the same NaCl samples as used for Fig. 4.34 and 4.37. X unannealed sample. ○, annealed sample.

The results of this type of analysis for the isotherm data illustrated [118] in Fig. 4.34 is shown in Fig. 4.37, in which the Hill-de Boer equation (4.72) was used as the isotherm function. This illustrates the growth of the main peak on annealing, which corresponds to the development of the (100) face of NaCl, and the decrease in the size of subsidiary peaks which are likely to correspond to higher order crystal planes.

Although the distribution obtained using the Fowler-Guggenheim equation for the isotherm function differs slightly, it is still of the same general shape, and has the major peak in virtually the same position on the energy scale. The isotherms for argon adsorption on the same samples gives a similar picture of what the process involved in annealing sodium chloride might be (see Fig. 4.38).

The chief limitations of the method at present appears to be the detailed experimental isotherms required, and the accuracy required. Even the data shown in part in Fig. 4.22 are not quite of sufficient accuracy to yield the information which should be determinable [84]. The methods are still undergoing development and refinement [119,120]. It should be remembered that an apparently good fit obtained with a particular isotherm function and resulting distribution function does not prove the correctness of both. It only indicates that the experimental isotherm can be accurately fitted by such a model. In view of the comments made earlier about hindered translation, and the fact that the model isotherm functions represent extremes of film mobility, no more should be envisaged.

4.6.7 Multilayer models

From early days of the determination of experimental physical adsorption isotherms, it was realised that adsorption normally did not stop at a monolayer and that multilayers were usually present when the relative pressure, $x = P/P_o$ (P_o is the saturation vapour pressure), is greater than about 0.1. Brunauer [121,122] noted that there seemed to be five different isotherm types, as illustrated in Fig. 4.39. These shapes are thought to have the following significance:

Type I is of the shape predicted by the Langmuir isotherm (4.77), being characterised by the assymptotic approach to the monolayer volume V_m^s, where $\theta = V^s/V_m^s$;

Type II is the most common type of isotherm corresponding to multilayer formation on a surface of high adsorption potential;

Type III is comparatively uncommon, corresponding to multilayer formation on a solid for which the monolayer adsorption potential is low, being of the same order as that for liquefaction of the adsorptive. Examples have been found for example for bromine on silica gel [122,123] and nitrogen on ice [124];

Types IV and V are the analogues of types II and III on porous adsorbents where the adsorption is limited by the volume of mesopores, causing the adsorption to level off at a pressure less than P_o. They reflect condensation phenomena and may show hysteresis effects.

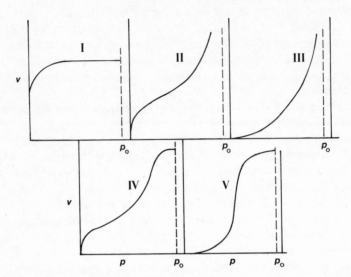

Fig. 4.39 — Brunauer's classification [121, 122] of five types of adsorption isotherm.

Although there had been a number of earlier attempts to extend the ideal localised monolayer treatment to include multilayer formation [105,125], it was not until 1938 that the Brunauer, Emmett and Teller [126] theory (BET) appeared in the literature. Although subsequently the BET isotherm has also been derived by statistical thermodynamic methods [127], the original kinetic derivation is still interesting and instructive.

The generalisation of the ideal localised monolayer treatment to include multilayer adsorption is achieved by assuming that each first layer adsorbed molecule serves as a site for the adsorption of a molecule into the second layer, and so on. The concept of localisation remains throughout all the layers and the forces of mutual interaction are neglected, as in the Langmuir isotherm. S_0, S_1, S_2, S_3 . . . . S_i. . . are defined as the areas covered by, respectively, 0, 1, 2, 3 . . . . i . . . molecular layers of adsorbate. At equilibrium the rate of condensation on S_0 is equal to the rate of evaporation from S_1, that is

$$a_1 P S_0 = b_1 S_1 e^{-E_1/RT} \tag{4.93}$$

where E_1 is the first layer heat of adsorption, assumed constant, and where a_1 and b_1 are constants. Equation (4.93) is formally identical with the simple Langmuir equation (4.76). Similarly the equilibrium between the first and second molecules is expressed by

$$a_2 P S_1 = b_2 S_2 e^{-E_2/RT} \quad . \tag{4.94}$$

and so on, the general statement of the equilibrium between the $(i-1)$th and ith layers being

$$a_i P S_{i-1} = b_i S_i e^{-E_i/RT} \quad .$$

(4.95)

The total surface area of the solid is given by

$$A = \sum_{i=0}^{i=\infty} S_i$$

(4.96)

and the total volume of adsorbed gas by

$$V^s = V_o^s \sum_{i=0}^{i=\infty} i S_i$$

(4.97)

V_o^s being the volume adsorbed per unit surface when the monolayer is completely filled. Dividing equation (4.97) by equation (4.96) gives

$$\frac{V^s}{A V_o^s} = \frac{V^s}{V_m^s} = \sum_{i=0}^{i=\infty} i S_i \Big/ \sum_{i=0}^{i=\infty} S_i$$

(4.98)

where V_m^s is the adsorbed volume in a complete ideal localised monolayer. Baly [125] had earlier arrived at what is essentially equation (4.98), but was able to proceed further only by empirical means. The key to the BET solution lies in the simplifying assumptions that

$$E_2 = E_3 = \cdots \cdot E_i = E_L$$

(4.99)

where E_L is the heat of liquefaction of the bulk liquid, and that

$$\frac{b_2}{a_2} = \frac{b_3}{a_3} \cdots \frac{b_i}{a_i} \quad .$$

(4.100)

In other words, the evaporation-condensation properties of the second and higher layers are assumed to be the same as those of the surface of the bulk liquid.

Equation (4.93) may be written

$$S_1 = y S_0$$

where

$$y = (a_1/b_1) P e^{E_1/RT}$$

(4.101)

Similarly, using equations (4.99) and (4.100) equation (4.94) may be written

$$S_2 = x S_1$$

where

$$x = (a_2/b_2) \, P e^{E_L/RT} \tag{4.102}$$

and in the general case

$$S_i = x S_{i-1} = x^{i-1} S_1 = y x^{i-1} S_0 = C x^i S_0 \tag{4.103}$$

where

$$C = \frac{y}{x} = \frac{a_1 b_2}{a_2 b_1} \, e^{(E_1 - E_L)/RT} \, . \tag{4.104}$$

Substitution of equation (4.94) in (4.98) yields

$$\frac{V^s}{V^s_m} = \frac{C S_0 \sum\limits_{i=1}^{i=\infty} i x^i}{S_0 \{ 1 + C \sum\limits_{i=1}^{i=\infty} x^i \}} \, . \tag{4.105}$$

The summation in the denominator is the sum of an infinite geometric progression given by

$$\sum_{i=1}^{i=\infty} x^i = \frac{x}{1 - x} \tag{4.106}$$

while that in the numerator may be transformed thus:

$$\sum_{i=1}^{i=\infty} i x^i = x \frac{d}{dx} \sum_{i=1}^{i=\infty} x^i = \frac{x}{(1 - x)^2} \, . \tag{4.107}$$

Substituting these results in equation (4.105) gives

$$\tag{4.108}$$

$$\frac{V^s}{V^s_m} = \frac{Cx}{(1 - x)(1 - x + Cx)}$$

On a free surface the amount adsorbed at saturation is infinite. Consequently at $P = P_0$, in order to make $V^s = \infty$, x in equation (4.108) must equal unity. From equation (4.102), therefore

$$(a_2/b_2) \, P_0 e^{E_L/RT} = 1 \tag{4.109}$$

when

$$x = P/P_0 \quad . \tag{4.110}$$

Substitution in equation (4.108) yields

$$V^s = \frac{V_m^s CP}{(P_0 - P)\{1 + (C - 1)P/P_0\}} \tag{4.111}$$

which is known as the 'simple' or '∞-form' BET equation. This equation may be transposed to the linear form

$$\frac{x}{V^s(1 - x)} = \frac{1}{V_m^s C} + \frac{C - 1}{V_m^s}\, x \quad . \tag{4.112}$$

Thus a plot of $x/V^s (1 - x)$ against x should give a straight line having slope $(C - 1)$ $(V_m^s\, C)$ and intercept $(1/V_m^s\, C)$. If the additional assumption is made that

$$\frac{a_1 b_2}{a_2 b_1} \approx 1 \tag{4.113}$$

the value of C is given by

$$C = e^{(E_1 - E_L)/RT} \quad . \tag{4.114}$$

The term $E_1 - E_L$, the difference between the heat of adsorption in the first layer and the heat of liquefaction, is known as the 'net' heat of adsorption. Thus the simple BET equation provides in terms of the model an estimate of both the heat of adsorption and the surface area of the solid. Further discussion of the determination of surface areas by the BET method will be found in section 4.6.8.

If, owing to lack of space, adsorption at saturation is restricted to n layers, the BET methods lead to the isotherm

$$V^s = \frac{V_m^s Cx}{(1 - x)}\, \frac{(1 - (n + 1)x^n + nx^{n+1})}{(1 + (C - 1)x - x^{n+1})} \tag{4.115}$$

where the quantities x, C and V_m^s have the same meanings as previously. Equation (4.115), which describes adsorption in a limited space such as a capillary, is commonly known as the 'n-layers' BET equation. This equation is of interest for two main reasons. Firstly it may be regarded as an empirical three constant isotherm equation, which has in the past proved useful in the evaluation of surface areas, and secondly, when $n = 1$ it reduces at all values of C (unlike the simple BET equation (4.129) which only reduces to this form when $x \ll 1$ and $C \gg 1$) to the Langmuir equation

$$V^s = \frac{V^s_m \, CP/P_o}{1 + CP/P_o} \tag{4.116}$$

which is the same as equation (4.79), where $\theta = V^s/V^s_m$ and $b = C/P_o$.

The simple BET equation (4.111) is capable of describing more or less adequately most Type II and Type III isotherms, the type being a function of the value of the constant C, but in general only type II isotherms corresponding to high C values yield reliable values of V^s_m.

The mathematical character of the BET equation can perhaps be more readily seen if it is rearranged in the form

$$\frac{V^s}{V^s_m} = \frac{1}{1 - x} - \frac{1}{1 + (C - 1)x} \tag{4.117}$$

where $x = P/P_o$. Thus the adsorption isotherm of V^s/V^s_m against x consists of the difference between the upper branches of two rectangular hyperbolae representing the first and second terms on the right-hand side of equation (4.117). The first has asymptotes at $V^s/V^s_m = 0$ and $x = 1$, while the second has asymptotes at $V^s/V^s_m = 0$ and $x = -1/(C - 1)$, and also cuts the V^s/V^s_m axis at $V^s/V^s_m = 1$. The difference between these two curves gives the BET isotherm, the shape of which is strongly dependent on C. For C values greater than 2 an isotherm of Type II results, whereas if C is less than 2 the resulting isotherm is of Type III. The point of inflection on the Type II curve is not necessarily at a value of $V^s/V^s_m = 1$ as has been often stated. It can be shown that this is the case when $C = 0$ and when $C \to \infty$. However, the discrepancy is only marked when $C < 9$. The effect of varying the constant C on the shape of the isotherm can be seen in Fig. 4.40.

There are certain problems associated with the application of the BET isotherm to experimental situations. Firstly it is usually found only to agree with experimental isotherms for non-porous heterogeneous surfaces in the pressure range $0.05 < x < 0.35$, whereas the model assumes a homotattic or homogeneous surface. The multilayer isotherm on such a uniform surface would be expected to be a series of steps [128], corresponding to the adsorption of successive layers, and these have been observed in practice; for example the adsorption of argon and krypton on graphitized P-33 carbon black at 77 K [129]. They are also partially visible for the adsorption of krypton on silver iodide referred to earlier [84], which can be thought of as a heterogeneous sample with some homotattic patches (Fig. 4.22).

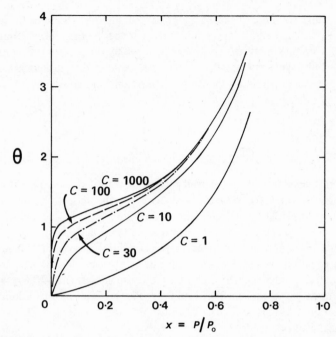

Fig. 4.40 – Theoretical BET isotherms (equation (4.111)) for various values of the constant C.

A number of attempts have been made to improve upon the BET 'simple' equation. Many of these are not well founded, but some still have importance. Anderson [130] assumed that the heat of adsorption in "the second to about the ninth layer" differs from E_L by a constant amount. This gives

$$\theta = \frac{C\,kx}{(1-kx)(1+(C-1)kx)} \qquad (4.118)$$

which is of the same form as (4.111) but with x multiplied by a constant, k. A second modification [131] allowed for the situation, likely to be found with porous solids, where the area available for adsorption decreases with successive layers. If the available area in the nth layer is j times that in the $(n-1)$th layer, the equation is

$$\theta = \frac{Cx}{(1-jx)(1+(c-j)x)}. \qquad (4.119)$$

As would be expected, increasing the number of parameters in the equations increases the range of relative pressure where the equations can be made to concur with experiment, but they give V_m^s values about 10% higher than the BET equation (4.111).

Again using the ideal localised monolayer model on a homotattic surface, Huttig [132] deduced a multilayer isotherm which assumes that the evaporation of a molecule from the ith layer is entirely unaffected by the presence of a molecule in the $(i + 1)$th layer, whereas the BET model assumes that the presence of an $(i + 1)$th molecule completely prevents evaporation from layers beneath it. This yields

$$\frac{x(1 + x)}{V^s} = \frac{1}{CV^s_m} + \frac{1}{V^s_m} x \quad .$$

(4.120)

Although this equation often fits well (ethane adsorbed on sodium chloride at 90 K fits equation (4.120) where $0.1 < x < 0.7$) it is theoretically badly founded, violating the principle of microscopic reversibility [133], and allowing the possibility of a molecule in the $(i + 1)$th layer being suspended in space with no molecule beneath it in the ith layer [134]. The values of V^s_m obtained by the Huttig equation are usually larger than those obtained by the BET method by an amount dependent on the C constant, but usually less than 20% greater.

An alternative approach to multimolecular adsorption was adopted by Harkins and Jura [135], in which they proposed the adoption of the equation of state for the adsorbed film

$$A = b - a\Pi$$

(4.121)

where a and b are constants, which is often an adequate representation of the liquid-condensed state of insoluble uncharged monolayers at the air-water interface. It would seem reasonable that a multilayer film of adsorbed gas could have similar properties. Applying the Gibbs transformation as in section 4.6.5 gives

$$\ln x = B - (A/(V^s)^2) \quad .$$

(4.122)

The specific surface area, s, may be related to the constant A by

$$s = k_{HJ} A^{1/2}$$

(4.123)

where k_{HJ} is a constant which has to be determined by the independent determination of s.

What is in many ways a very attractive method of visualising multilayer adsorption is to examine the parallel between the distribution of the gas molecules of the earth's atmosphere in the earth's gravitational field, and adsorbed gas molecules at the surface of a solid in its adsorption potential field (Fig. 4.41). These ideas were first set down by Polanyi [136], who envisaged equipotential surfaces, the distance between which correspond to definite volumes, and that there will be a relationship between the potential ϵ and the volume ϕ. Thus the adsorption process may be represented by the function $\epsilon = f(\phi)$, which is a distribution function. Since the potential is independent of temperature, then $\epsilon = f(\phi)$ represents a 'characteristic curve' for adsorption at all temperatures.

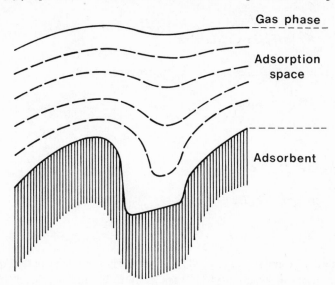

Fig. 4.41 – Cross-section of the adsorbed layer showing equipotential contours according to potential theory (After Brunauer [121]).

Polanyi postulated three separate cases describing the state of the adsorbed film:

Case I At temperatures well below the critical temperature of the adsorbate, the adsorbed film will be a liquid;

Case II At temperatures only just below the critical temperature the adsorbed layer will be a mixture of liquid and compressed gas;

Case III At temperatures above the critical temperature the adsorbed phase is a compressed gas.

Berenyi [116, 137] elaborated the treatment of Case I by assuming that the gas phase is ideal, the liquid adsorbed film is incompressible, and that a negligible amount of work is involved in forming the liquid surface. He showed that the potential at a distance x from the surface is given by

$$\epsilon_x = RT \ln(P_o/P) \tag{4.124}$$

where P is the equilibrium pressure, as previously.

Considering multilayer adsorption only, if we make the assumption that only dispersion forces are responsible for adsorption, then the potential should decrease with the inverse cube of the distance. Then

$$\epsilon_x = \epsilon_0/(a+x)^3 \tag{4.125}$$

where ϵ_0 is a constant [113] and a is a distance defining the closest approach of the adsorbate molecule to the surface. Combining equation (4.124) and (4.125) gives

$$RT \ln (P_0/P) = \epsilon_0/(a+x)^3 \tag{4.126}$$

from which the isotherm equation may be deduced

$$V^s = -\alpha + \beta \, \omega^{-1/3} \, , \tag{4.127}$$

where

$$\alpha = a \, v_0 \, S/v_1$$

$$\beta = (\alpha/a)(\epsilon_0/RT)^{1/3} \text{ and}$$

$$\omega = \ln(P_0/P) \quad ,$$

in which v_0 is the molar volume at STP and v_1 is the molar volume of the liquid. The chief limitations of equation (4.127) are likely to be that at low values of P/P_0 where the adsorbed film is thin, then the assumption that the solid is continuous, which is the basis of equation (4.125), is likely to be a poor one. Furthermore if on ionic solids the electric field contribution to the adsorption potential is appreciable then equation (4.125) will once again be inappropriate. This approach which is due to Frenkel [138], Halsey [139] and Hill [140], is more commonly quoted in the form where $a = 0$, which gives

$$\theta^n = A_{FHH}/\ln (P_0/P) \tag{4.128}$$

where

$$A_{FHH} = \epsilon_0(x_m)^n RT$$

and x_m is the adsorbed film thickness at the monolayer point. This equation is frequently called the Frenkel-Halsey-Hill equation, and their theory referred to as Slab Theory. Experimental values of n, which should be 3 if van der Waals dispersion forces alone are responsible for adsorption, are frequently [141] less than 3, which is interesting since a value of $n = 2$ gives a form of equation (4.128) equivalent to the Harkins-Jura equation (4.122).

An alternative form for ϵ_x was proposed by de Boer and Zwikker [142] based on the assumption that a polar adsorbent surface could induce dipoles in non-polar adsorbate molecules in the first layer, and that these in turn could induce dipoles in the second layer, and so on. This would predict the form of equation

$$\epsilon_x = \epsilon_o \exp\left(-\chi x\right) \qquad (4.129)$$

where

$$\chi = -(1/a_o)\ln(\alpha/a^3)^2 ,$$

in which α is the polarisability, a_o is the atomic diameter, and a the 'effective' value of distance from the surface, which is smaller than a_o. Combining this with equation (4.124) and solving as before gives the equation

$$\ln\ln\left(P_o/P\right) = \ln\left(\epsilon_o/RT\right) - (\chi v/v_o S)\,V^s \qquad (4.130)$$

which is a suitable linear form for the isotherm.

It is important to realise that all the above multilayer models assume a uniform surface, whereas all surfaces are to a degree heterogeneous, and most are very heterogeneous. However, as we shall see in the next section, this has not prevented the application of these models to experimental isotherms on heterogeneous surfaces. The only significant attempt to tackle this problem is that due to Halsey [143], who extended the application of equation (4.127) to heterogeneous surfaces, with $n = 3$, using a 'patchwise' model. The method is based on distribution function $N_1(V_o)$, giving for the coverage in the mth layer

$$\theta_m = \frac{1}{m^3} \int_{-m^3\,RT\,\ln(P/P_o)}^{\infty} N_1(V_o)\mathrm{d}U_o \qquad (4.131)$$

which leads to the isotherm equation

$$\theta = \exp\left(U_o/\bar{U}_o\right) \sum_{m=1}^{\infty} z^{m^3} \qquad (4.132)$$

where $z = (P/P_o)^{RT/U_o}$, and $\bar{U}_o$ is the modulus of the distribution. Although this equation is difficult to apply to experimental isotherms it does predict a stepped isotherm for a uniform surface, with the gradual disappearance of the steps as the degree of heterogeneity increases.

4.6.8 Surface area determination

In the previous section we discussed the various models for the multilayer adsorption processes. Firstly therefore we should examine the applicability of these to experimental isotherms, and see what sort of 'goodness of fit' we might expect. If we use readily accessible data, then the reader can extend the process further if he wishes. Data which has been previously used for this purpose [144] is that of Keenan and Holmes [145] who examined the adsorption of nitrogen, argon and oxygen at several temperatures on potassium chloride. We will use two of the most extensive isotherms, namely nitrogen at 78K and argon at 83K.

These two isotherms are both BET type II, which as we have said before is the most common multilayer form, and are shown in Fig. 4.42. These two curves have the appearance of one with a fairly high C constant (see Fig. 4.40) in the case of N_2, and one with a low C constant for the Ar isotherm. On the latter isotherm the linear portion at low pressure which does not extrapolate back to the origin looks slightly abnormal; a linear portion below the pressure range of these data would correspond to Henry's Law (see section 4.6.5). Argon and nitrogen are the two most frequently recommended gases for surface area determination.

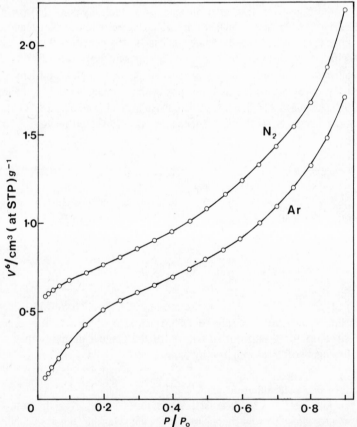

Fig. 4.42 – Adsorption isotherms of nitrogen at 78 K and argon at 83 K on potassium chloride. Data of Keenan and Holmes [145].

Let us examine the application of the BET 'simple' model to the data. The conventional BET plot is shown in Fig. 4.43 using equation (4.112). The N_2 data gives a linear plot over the approximate range $0.03 \leqslant x \leqslant 0.25$, whereas the linear region of the Ar data is difficult to see, but the range $0.07 \leqslant x \leqslant 0.15$ does seem to be linear. Since both isotherms were determined on the same sample then one would like to find that the surface area calculated from both plots is the same.

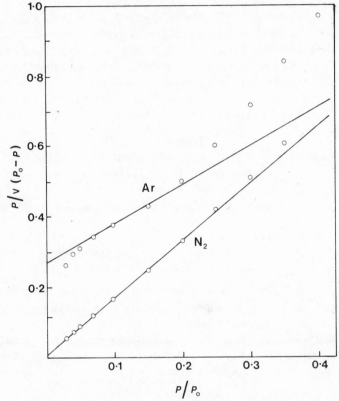

Fig. 4.43 – BET plots of the adsorption data of Keenan and Holmes [145] shown in Fig. 4.42.

From the linear region a value of V_m^s can be obtained from

$$V_m^s = \frac{1}{\text{slope} + \text{intercept}} = \frac{1}{\dfrac{C-1}{CV_m^s} + \dfrac{1}{CV_m^s}} \qquad (4.133)$$

and a value of the BET constant from

$$C = 1/(1 - \text{slope} \times V_m^s) \quad . \qquad (4.134)$$

In order to calculate the specific surface area we must use the equation

$$s = \frac{V_m^s L \, \sigma_m}{v_0} \quad m^2 g^{-1}$$

$$= 26.87 \, \sigma_m V_m^s \quad m^2 g^{-1} \tag{4.135}$$

if V_m^s is in cm^3 at STP g^{-1} and σ_m in nm^2 $molec^{-1}$, and v_0 is the molar volume. The quantity σ_m is known as the 'molecular cross-sectional area of the adsorbate' or better as the 'adsorption cross-section', and is the area occupied by an adsorbate molecule in a complete monolayer. It is usually calculated from the liquid density of the adsorbate at the adsorption temperature and may be taken as 0.162 nm^2 $molec^{-1}$ and 0.138 nm^2 $molec^{-1}$ for N_2 and Ar respectively [146]. The results of the BET analyses are shown in Table 4.8.

Table 4.8
BET analysis of the plots in Fig. 4.43 of the isotherm data shown in Fig. 4.42 of Keenan and Holmes [145].

	Slope	Intercept	V_m^s/cm^3 at STP g^{-1}	C	$\sigma_m/nm^2 \, molec^{-1}$	s/m^2g^{-1}
N_2 at 78K	1.620	0.006	0.615	270	0.162	2.68
Ar at 83K	1.108	0.262	0.730	5.2	0.138	2.71

The values shown in this table indicate close agreement in the value of the specific surface area, and the expected diversity of C constant values. Thus despite the problems, consistent results are obtained.

The modifications to BET theory present various problems. Firstly the Anderson equations (4.118) and (4.119) do not have readily plottable forms, unless it is assumed that C is high [147], when

$$x = \frac{1}{k} \left[1 - \frac{V_m^s}{V^s} \right] . \tag{4.136}$$

It would be expected that this simplification might be justified for the N_2 data but not for the Ar data. The plot of this equation is shown in Fig. 4.44, and contains the surprise of an apparently quite good fit for the Ar data over the range $0.3 \leqslant x \leqslant 0.9$ compared with the more limited range found with the N_2 data of $0.2 \leqslant x \leqslant 0.6$. The linear sections do not give consistent surface areas since $s_{N2} = 3.05$ m^2g^{-1} and $s_{Ar} = 2.54$ m^2g^{-1}. Thus the least valid case gives the longer fit range and the specific surface area nearer the BET answer, which offers another example of the fact that an apparent 'fit' of the data does not prove the validity of the model.

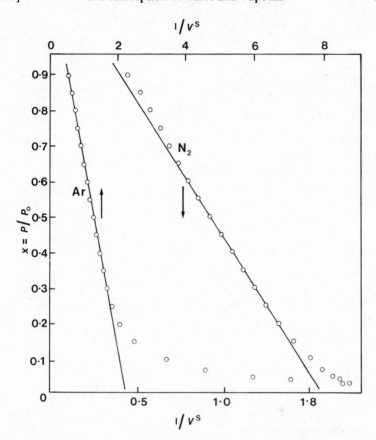

Fig. 4.44 — Plot of the reduced Anderson equation (4.136) for the adsorption data of Keenan and Holmes [145] shown in Fig. 4.42.

The Huttig equation (4.120) is another example of reasonable 'fits' (Fig. 4.45) giving inconsistent results, and since the model upon which it is based is erroneous, nothing much can be gained from the answers that $s_{N_2} = 2.85$ m^{2}g^{-1} and $s_{Ar} = 2.18$ m^{2}g^{-1}.

Plotting the data using the axes of the linear form of the Harkins-Jura equation (4.122) gives the graphs shown in Fig. 4.46. The shape of the plot for the N$_2$ data is that usually found with two linear regions, which has been suggested is indicative of there being two liquid-condensed states of the adsorbed film, and Harkins and Jura [135,148] have suggested that the lower pressure phase be used to calculate s. This cannot be done in the present case because of the absence of reference data on a standard potassium chloride sample which would be needed to evaluate the constants. The situation can be even more complicated since it has been suggested [149] that there may be four different liquid phases in the case of N$_2$ on carbon black.

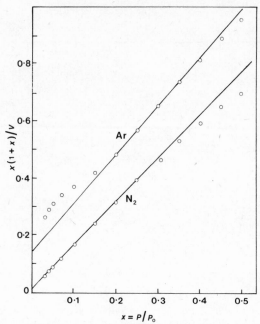

Fig. 4.45 — Huttig plots, equation (4.120), of the adsorption data of Keenan and
Holmes [145] shown in Fig. 4.42.

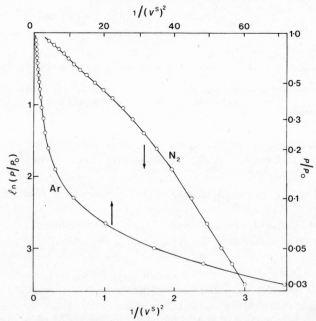

Fig. 4.46 — Harkins-Jura equation (4.122) plots of the adsorption data of Keenan
and Holmes [145] shown in Fig. 4.42.

The argon data in Fig. 4.46 show only one linear region, and would tend to suggest that a liquid condensed state model in this case is only applicable when coverages are greater than about a monolayer. All the N_2 data presented would be in this region. The low pressure N_2 data given by Harkins and Jura [135] also tend to curve in a manner similar to the present Ar data.

The remaining two models worth discussing are potential theory using equation (4.127) and plotted in Fig. 4.47, and polarisation theory, using equation (4.130) and plotted in Fig. 4.48. Both models represent the data reasonably well, but potential theory is the more successful in this respect. However, there are serious difficulties in evaluating the constants to permit their use in the estimation of surface area although the models could be used in a relative fashion, as was the case with the Harkins-Jura equation.

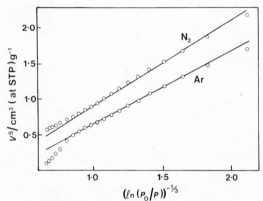

Fig. 4.47 — Potential theory plots using equation (4.127) of the adsorption data of Keenan and Holmes [145] shown in Fig. 4.42.

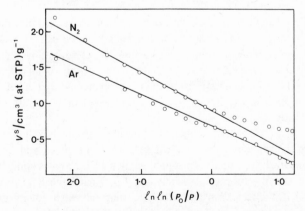

Fig. 4.48 — Polarization theory plots using equation (4.130) of the adsorption data of Keenan and Holmes [145] shown in Fig. 4.42.

The above example highlights the fact that the BET 'simple' equation offers the most practically useful method of evaluating surface areas. The remainder of this section will therefore consider the three gases most frequently used in this connection namely N_2, Ar, and for small specific surface areas Kr.

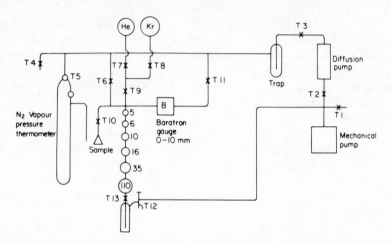

Fig. 4.49 – Schematic diagram of a design for a krypton BET volumetric adsorption apparatus [151]. (Reproduced with permission from Jaycock [151]).

The techniques of measuring the amount of gas adsorbed at a particular equilibrium pressure are usually volumetric and have been reviewed quite frequently [150], and the schematic diagram of an apparatus for krypton BET measurements [151] is shown in Fig. 4.49. Although the standard apparatus has been proposed for the nitrogen BET method, the problem is that the actual design is in some measure a matter of personal preference, and also frequently a function of the amount of money available as well as the availability of technological innovations. However, it is useful for anyone entering the field for the first time to be able to see a sample design, which can be varied to suit circumstances, provided that simple calculations are performed to check on the effects of such changes on accuracy. Because the design of an apparatus for the krypton BET method illustrates most of the problems that are likely to be encountered with any gas, this is the type illustrated in Fig. 4.49, and which will now be discussed in more detail.

Fig. 4.49 shows a schematic diagram of a design based on the following assumptions: (i) that the sample has an area 0.1 m^2, (ii) that $C \sim 35$, and (iii) that the minimum volume of the system is not greater than 15 cm^3. This volume is the sum of the dead space, the tube volume between T_{10}, T_9 and the Baratron, and the Baratron gauge itself. These assumptions mean that only when they are complied with will the apparatus give 6 points, spread throughout the BET range, from one dose of krypton. Deviations from these assumptions may result in a reduction of the number of points within the range, or a less even spread, but for many samples they will prove satisfactory. Clearly the design represents a compromise which the authors believe to be satisfactory.

There are a number of specific experimental problems which should be understood and which will be dealt with individually.

(a) *Pressure-measuring devices.* The design shows a 0–10 torr Baratron diaphragm pressure gauge which is quick to use, and is sufficiently precise to make pressure measurement an insignificant source of error. Simpler devices such as McLeod gauges or thermistor gauges can be used, but they result in a loss of accuracy in pressure measurement. The use of a McLeod gauge with a system such as that in Fig. 4.49 would necessitate changes in the gas burette bulb sizes, since the minimum system volume is likely to be ~150 cm^3.

(b) *Dead space calibration.* This is normally done by calibration with helium and results in an equivalent volume, V, at room temperature, T_R. This volume is dependent on T_R.

Let us suppose that V arises from two components, a volume, V_R, above the liquid nitrogen level at T_R, and an actual physical volume, V_N, below liquid nitrogen level at liquid nitrogen temperature, T_N. The total number of moles in the system may then be written

$$n = \frac{PV}{RT_R} = \frac{PV_R}{RT_R} + \frac{PV_N}{RT_N} \quad . \tag{4.137}$$

Therefore

$$V = V_R + V_N \frac{T_R}{T_N} \quad . \tag{4.138}$$

It is essential, therefore, either to know the values of V_R and V_N or to only use a value of V appropriate to the temperature at the time adsorption measurements are made. Normally errors of this type are small since T_R does not usually change greatly, but, for example, intermittent sunshine can cause problems. Errors due to too large a dead space have been discussed by Ross and Olivier [152]. In general it is undesirable that the dead space should be larger than the burette volume.

(c) *Effect of light.* It has been noted by Rosenberg and Martel [153] that small amounts of light falling on the sample can result in a significant rise of the sample temperature. The sample should therefore always be protected from light and other radiation. Cover glass sample tubes with copper foil, that has been blackened with yellow ammonium sulphide, up to the liquid nitrogen level, and cover the tube with black paint from this junction to beyond the tap isolating the sample from the rest of the system. The copper foil should not extend above the liquid nitrogen level since it would spoil the sharpness of the thermal junction.

(d) *Thermal transpiration.* This effect causes a reduction of the actual pressure compared to that measured on the gauge, at the lower end of the pressure range. An idea of the errors involved can be seen in the work of Ross and Olivier [154]. Using a 6 mm internal diameter sample tube, they suggest the error would be ~2% at $P = 0.12$ torr or $P/P_o = 0.05$, and that it decreases as the sample pressure is increased. The effect would be greater if the sample tube diameter at the junction were less than 6 mm. A 2 mm diameter tube would result in an increase in the error to about 10% at the same pressure. Calculations should normally be performed to see whether the error is significant, and the most suitable recent data for this purpose are those of House and Jaycock [155].

If corrections are to be made then the following points need to be considered. Firstly the equilibrium pressure will be the corrected one, and secondly the correct volumetric estimation of the amount adsorbed involves recognising that there are two pressures in the dead space volume, and that the number of moles of gas present will be given by

$$n = \frac{P_R V_R}{RT_R} + \frac{P_N V_N}{RT_N} \qquad (4.139)$$

where P_R and P_N are the pressures in the parts of the system at T_R and T_N respectively, and V_R and V_N need to be known. These two volumes may be calculated from measurements of the 'dead space' first with liquid nitrogen present, and then with the whole system at room temperature.

(e) *Measurement and control of sample temperature.* Sample temperature can best be measured by means of a nitrogen saturation vapour pressure thermometer using standard vapour pressure data [156]. This permits the bath temperature to be measured to at least 0.01 K. The temperature can be controlled by an expensive cryostat, but can be simply done using mixtures of liquid nitrogen and liquid air. If a pressure of about 770 torr on the vapour pressure thermometer is chosen (~77.4 K) then the bath temperature can be fixed at this by adding small amounts of liquid nitrogen or air to allow for atmospheric pressure variation. The system should be stirred using a slow-speed electric motor to ensure uniformity of temperature, and the liquid nitrogen/air level on the sample tube should be kept constant.

There is an alternative method for observing the temperature, although it does not permit continuous control, and only offers a single measurement at the end of the adsorption measurements. After the adsorption measurements are complete a large dose of krypton can be admitted to the system and the saturation vapour pressure of solid krypton measured. This may be used to calculate the temperature using the vapour pressure data of Keesom *et al* [157]. It is preferable to use a nitrogen vapour pressure thermometer.

(f) *The non-ideality of krypton.* Calculations show that with pressures less than 2 torr no correction is necessary for non-ideality, the correction terms being

(1-0.00229 P) at room temperature and (1-0.176 P) at liquid nitrogen temperature, where P is in atmospheres. However, corrections may be significant if a McLeod gauge is used to measure the pressure, but in the system in equilibrium with the solid the error in neglecting non-ideality correction would be a maximum of 0.04% at 77.5 K and 0.0005% at room temperature.

The above considerations are all relevant to N_2 and Ar measurements with the exception of thermal transpiration corrections, which are not necessary in conventional apparatus in the pressure ranges where $x > 0.03$.

If Kr is used there also remains the problem of the appropriate value of P_o. The adsorbed film in fact appears to behave more like a liquid than a solid [158], and therefore a vapour pressure equation for the liquid state needs to be extrapolated [151], since Kr is a solid at liquid N_2 temperatures.

In order to convert V_m^s values into specific surface areas, then a value for σ_m has to be known. The values usually quoted are calculated from liquid densities, and therefore assume that the adsorbed molecules pack in a similar way on the surface [158]. The fact that the adsorbed film is a liquid is not surprising from the point of view of potential theory, but the hindered translation model calculations discussed earlier would suggest that some degree of localisation will be met with quite frequently, and that the effective value of σ_m for the first layer will be larger than that calculated from the liquid density in this case. This point may be clearly seen in relation to the values σ_m frequently employed in the calculation of surface areas by the krypton BET method, and some are listed in Table 4.9.

Table 4.9
Some of the proposed values of σ_m for krypton at 77.5K

Method of calibration	σ_m/nm^2 molec^{-1}	Reference
Standard anatase	0.195	[158]
	0.218	[159]
N_2 BET at 77.5K	0.195	[158]
$\sigma_m = 0.162$ nm^2 molec^{-1}	0.208	[160]
	0.218	[159]
	0.226	[161]
	0.236	[161]
Liquid density	0.152	[158]
Solid density	0.140	[158]

This table clearly shows that the necessary experimental values of σ_m are significantly larger than those predicted from either liquid or solid densities, which is consistent with some degree of localisation of the adsorbed film.

The BET equation is normally applicable in the relative pressure range between 0.05 and 0.35, or some slightly more restricted range within those limits. It is often stated that this corresponds to the pressure range where the second layer of adsorbed molecules is being built up on the surface. In many cases krypton BET plots give values of the BET constant, C, which are less than 100. Table 4.10 shows how the coverage at the two extremes of the BET range, and the coverage at the statistical monolayer point varies with C. It can be seen that when $C > 200$ then the BET range corresponds approximately to $1 < \theta < 1.5$, and that when $C < 5$ to $\theta < 1$. However, in the range of C values frequently found for krypton adsorption, $10 < C < 100$, the statistical coverage will vary from first layer to half-filled second layer. In the light of this it is worth considering the relative molecular packing and the adsorption forces in the first and second layers. The molecular packing may well be different in the first and second layers, and therefore the value of σ_m used in BET calculations will be an average of some sort that reflects this.

Table 4.10

The values of the fractional coverage, θ, at relative pressure values $x = 0.05$ and $x = 0.35$, and the value of x when $\theta = 1$, as a function of the BET constant, C.

C	θ at $x = 0.05$	θ at $x = 0.35$	x when $\theta = 1$
2	0.100	0.798	0.415
5	0.219	1.122	0.309
10	0.363	1.297	0.241
30	0.644	1.449	0.155
60	0.799	1.492	0.115
100	0.855	1.510	0.091
200	0.961	1.524	0.066
500	1.014	1.533	0.043

Singleton and Halsey [162] and Bonnetain et al [163] have argued that because of localisation or pseudo-localisation of the molecules in the first layer, the quasi-lattice so formed influences the stacking of molecules in the second layer and that the multilayer steps such as those found by Thomy and Duval [164], are only found when the adsorbate and adsorbent crystal lattices are geometrically compatible. On this basis Gregg and Sing argue that the application of the BET procedure to any gas at temperatures below its triple point is questionable [165]. However, problems associated with localisation, or perhaps what is better called hindered translation, are much more general than that, and krypton merely represents another awkward example.

The spacing of the adsorbed atoms in a localised film will be entirely determined by the adsorption potential well distribution, which is in turn determined by the surface structure of the adsorbate. This must of necessity vary from substance to substance and will also be affected in some measure by surface contamination and surface defect concentration. In this case a universal figure for the molecular adsorption cross-section is clearly unattainable, but it may be possible to assign unambiguously a value for a particular adsorbent, although this has not been achieved so far.

For a truly mobile film it is probably reasonable to assume, particularly in the light of the other assumptions in the BET theory, that the packing may approximate to that in the corresponding bulk phase, be it liquid or solid. Emmett and Brunauer [166] showed the general relationship would be

$$\sigma_m = f \left[\frac{M}{\rho L} \right]^{2/3} \times 10^{14} \, nm^2 \, molec^{-1} \tag{4.140}$$

where M is the molecular weight, ρ the density of the appropriate phase and L Avogadro's Number. The packing factor, f, depends on the geometry of the adsorbed phase and equals 1.091, for hexagonal close packing. Calculations based on the liquid and solid densities yield values which must represent lower limits for σ_m, since one cannot envisage atoms closer packed than in the solid state.

That some degree of localisation is apparently found with the majority of gases can be seen in Table 4.11 where experimentally required values of σ_m exceed those predicted by equation (4.140) except when the likely restriction of rotation in the adsorbed film of a non-spherically symmetric molecule, can cause an increase in packing density. In this table the assumption is being made that an adsorbed N_2 film is totally mobile.

Table 4.11

Experimental values of σ_m for various gases compared with values derived from the liquid density.

Gas	Temperature/K	$\sigma_m/nm^2\ molec^{-1}$		Reference
		Eqn. (4.140)	Exptl.	
N_2	77.5	0.162	0.162	[167]
Ar	77.5	0.128	0.169	[167]
Kr	77.5	0.152	0.195	[158]
CO	77.5	0.160	0.147	[159]
C_5H_{12}	293	0.362	0.523	[167]
C_6H_{14}	273	0.390	0.589	[168]
C_7H_{16}	298	0.440	0.640	[168]

4.6.9 Porosity

The investigation of many porous solids involves the use of gas adsorption methods. Pores are usually classified according to size in the manner originally proposed by Dubinin [169]. Pores having width less than 2 nm are termed *micropores,* those between 2 and 50 nm, *mesopores,* and those whose width is larger than 50 nm, *macropores.* The first two classifications can be investigated by gas adsorption methods, but the latter appear too similar to a plane surface for adsorption differences to be significant, but they can be investigated by the measurement of the pressure required to force mercury into them, and to convert the volume of mercury forced in at a particular pressure into a pore volume vs. pore radius plot using the Laplace equation (3.10).

One approach to adsorption in micropores is by means of potential theory, and was developed by Dubinin *et al* [170-172]. They assumed that for two different adsorptives filling the micropore adsorption space to the same extent, W, the adsorption potentials will possess a constant ratio to one another, thus

$$\frac{\epsilon_1}{\epsilon_2} = \beta \qquad (4.141)$$

where β is termed the affinity coefficient. If adsorbate 2 is chosen as an arbitrary reference standard, then we may write equation (4.141) as

$$\frac{\epsilon}{\epsilon_0} = \beta \qquad (4.142)$$

where the subscript o denotes the standard adsorbate. For this standard we may write

$$W = f(\epsilon_0) = f(\epsilon/\beta) \quad . \qquad (4.143)$$

Now if W is a Gaussian function of the adsorption potential, then

$$W = W_0 \exp\left(-\tilde{k}\ \epsilon_0^2\right) \qquad (4.144)$$

where $\tilde{k}$ is a constant characterising the pore distribution. Using equations (4.124) and (4.142), and also using the fact that $W = n^s/\rho$ where ρ is the molar density of the liquid adsorbate, then

$$\ln n^s = \ln(W_0\rho) - \frac{\tilde{k}\ R^2 T^2}{\beta^2}\ \{\ln (P_0/P)\}^2 \qquad (4.145)$$

which is known as the Dubinin-Radushkevich equation, where W_o is the total micropore volume. Data [173] for an activated carbon made from sucrose for the adsorption of benzene at 293K are shown in Fig. 4.50, plotted according to equation (4.145).

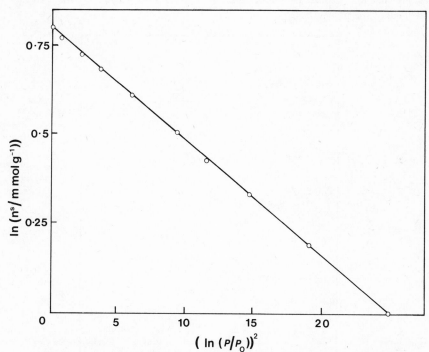

Fig. 4.50 − Plot of the data [173] for the adsorption of benzene at 293 K on an activated carbon made from sucrose, using the Dubinin-Radushkevich equation (4.145).

Kaganer [174] has pointed out that an equation similar in form would be found in the monolayer region for a planar surface with a Gaussian distribution of adsorption potentials. Thus if we write

$$\theta = \exp\left(-\tilde{k}_1 \epsilon^2\right), \tag{4.146}$$

then equation (4.144) has the analogous form

$$\ln n^s = \ln n^s_m - \tilde{k}_1 R^2 T^2 \{\ln (P_o/P)\}^2 \quad . \tag{4.147}$$

Thus for a microporous solid the value of the micropore volume may be obtained from the intercept on the plot of $\ln n^s$ vs$(\ln (P/P_o))^2$, or the monolayer capacity, n^s_m, from the same plot for a non-porous material, if the assumptions are plausible.

Such plots are not always easy to interpret as can be seen in Fig. 4.51 for the adsorption of krypton on a tin (IV) oxide sample at 77.5K. This sample is thought to be partially microporous, and it is possible that the low pressure region is primarily controlled by adsorption in micropores, and that upturn from this line is due to mesopores or multilayer formation on the external surface of the porous body.

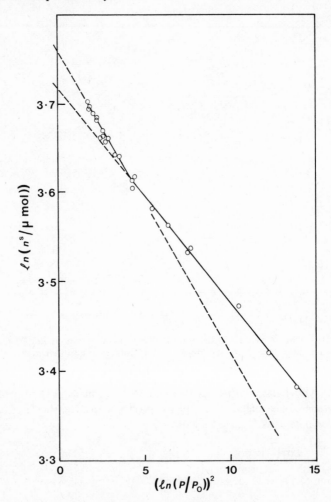

Fig. 4.51 – Dubinin-Radushkevich plot for the adsorption of krypton at 77.5 K on tin (IV) oxide.

It should be noted that if the adsorbent is entirely microporous a Type I isotherm will be experimentally determined, and this may sometimes be approximated by Langmuir's equation (4.77). It has been shown [175] that an almost linear plot results using Dubinin-Radushkevich axes when $b = 18/P_o$ at pressures where $P/P_o < 0.25$, and that the graphs are appreciably curved at both higher and lower b values.

The effects of mesopores on the shape of an adsorption isotherm is illustrated diagrammatically in Fig. 4.52, where the transition from a Type II isotherm, A, to a Type IV isotherm, B is shown. The two separate branches of the B isotherm correspond to the adsorption and desorption processes, that is to say, the isotherm shows hysteresis. Because of capillary condensation in the mesopores, the uptake at a particular relative pressure will be increased by the amount of condensate in those pores. If pores of a suitable shape are present, then it is possible for capillary condensation to occur below the pressure where the hysteresis loop starts. The origin of the hysteresis loop has been the subject of considerable debate. There are essentially two primary theories. The first is that the radius of curvature of the liquid surface in the capillary differs as the capillary fills from that as it empties as a result of differences normally expected between advancing and receding contact angles. This concept, due to Zsigmondy [176], has always been open to the objection that these angles are not susceptible to direct measurement.

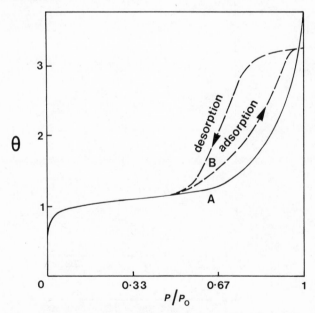

Fig. 4.52 – Diagrammatic representation of a Type IV isotherm for adsorption on a mesoporous adsorbent.

The alternative ideas [177,178] assume that hysteresis is due to 'ink bottle' shape pores, such that the wider cross-section of the pores has to be filled before the narrow one, and that during desorption the reverse occurs. If the contact angle is assumed to be zero, then the radius of the pore may be calculated by means of Kelvin's equation (3.8), where the radius in this equation, r_k, the Kelvin radius, is related to the pore radius, r_p, by

$$r_p = r_k + t \qquad (4.148)$$

where t is the adsorbed film thickness. In order to evaluate r_p, then values of t must be obtained from a standard isotherm, either experimental or theoretical.

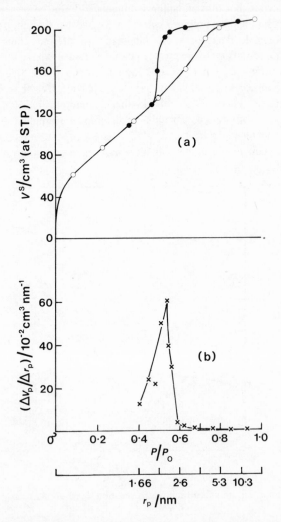

Fig. 4.53 – Pore size distribution from the nitrogen adsorption isotherm at 77.5 K on titanium (IV) oxide. Data of Harris and Whitaker [181]. (a) adsorption isotherm; ○, adsorption; ●, desorption. (b) Pore size distribution curve. V_p and r_p are the pore volume and radius respectively.

Pierce [179,180] has carefully described a calculation scheme for carrying out the derivation of a pore size distribution from a Type IV isotherm. The results of one such calculation [187] are shown in Fig. 4.53.

An alternative way of analysing nitrogen adsorption data at 77.5K on silica and alumina adsorbents in terms of their pore structure was proposed by de Boer *et al* [182,183], but it requires the use of a standard isotherm determined on a non-porous sample of the adsorbent. This isotherm is represented not as $\theta = f(P/P_o)$, but as $t = f(P/P_o)$, where t is the statistical thickness of the adsorbed layer calculated using the identity that when $\theta = 1$, $t = 0.354$ nm. Therefore if the experimental isotherm $V^s = f(P/P_o)$ is determined, then a value of t can be found from the standard isotherm, using a particular P/P_o value. The pairs of values of V^s and t may then be plotted as illustrated in Fig. 4.54.

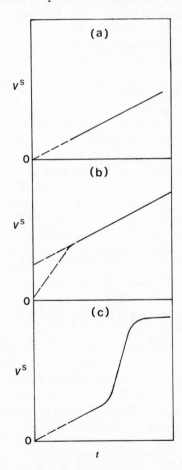

Fig. 4.54 – Diagrammatic representation of de Boer '*t*-plots' for (a) – non-porous surface, (b) – with some micropores, and (c) – with some mesopores.

If we consider Fig. 4.54(a) for a non-porous surface, then the slope will be proportional to the surface area. If t is expressed in Å and V^s in cm^3 at STP g^{-1}, then the specific surface area may be calculated from

$$s = 15.47 \times \text{slope m}^2\text{g}^{-1} \quad . \tag{4.149}$$

Since the standard isotherm of de Boer covers only the region where $P/P_o > 0.08$, the data are confined to the multilayer region only ($\theta = 1.0$ at $P/P_o = 0.08$), the effects of micropores are usually only experimentally observable at the lowest pressure points of experimental isotherms, and only a few are in the standard isotherm range. Fig. 4.54(b) shows a t-plot for a microporous surface with the same external surface as in (a), as indicated by the second slope, but with micropores which fill first. The initial slope does not really represent a total surface area, since surface area is a concept which has little real significance for a micropore. However, extrapolation of the higher pressure slope back to the V^s axis gives an intercept proportional to the amount in the micropores, which can be related to the micropore volume by

$$W_o = V^s_{\text{intercept}}/22\,414\,\rho \quad . \tag{4.150}$$

Fig. 4.54(c) shows the t-plot for a solid with the same total surface as in (a), as indicated by the initial slope, but the effects of capillary condensation cause a sudden upturn in the graph, and the final very low slope corresponds to the external surface of the adsorbent which is still available after the capillaries are full. Combinations of types (b) and (c) are not unknown.

Sing [184,185] has introduced a variation of the 't-plot' known as the 'α_s-plot' which essentially avoids using a monolayer capacity in the definition of the standard isotherm, which is instead expressed as V^s/V^s_s, where P/P_o has some arbitrarily selected value. In practice Sing usually employs $P/P_o = 0.4$, where it is argued any micropores would have been filled, and capillary condensation has not started. This is useful if the standard isotherm does not possess a sharp knee, and in consequence methods of evaluating V^s_m are likely to be somewhat inaccurate.

Perhaps the most convincing proof of the validity of pore volume analysis lies in what is known as Gurvitsch's rule, which is that although the amount adsorbed at saturation, n^s or V^s, varies from adsorptive to adsorptive, it represents a constant volume when expressed as a volume of liquid, v_1. McKee's results [186] for silica gel at 25°C are listed in Table 4.12. The divergence in v_1 is only slightly greater than experimental error.

Table 4.12

McKee's results [185] for the uptake at saturation, n_{sat}^s, expressed as a volume of liquid, v_1, for various adsorbates on silica gel at 25°C.

Adsorbate	$v_1/cm^3 g^{-1}$	n_{sat}^s/m mol g^{-1}
Hexane	0.431	3.28
2,3-Dimethylbutane	0.429	3.28
2-Methylpentane	0.431	3.28
Heptane	0.431	2.91
2,2,3-Trimethylbutane	0.420	2.88
Octane	0.434	2.66
2,2,4-Trimethylpentane	0.439	2.63
2,3,4-Trimethylpentane	0.425	2.66
Cyclohexane	0.421	3.88
Methyl cyclohexane	0.425	3.32
Ethyl cyclohexane	0.426	2.99
Benzene	0.440	4.92
Nitromethane	0.449	8.33
Nitroethane	0.434	6.03
Carbon tetrachloride	0.421	4.30

4.7 REFERENCES

[1] R. Shuttleworth, (1950) *Proc. Phys. Soc. (London)*, **A63**, 444.

[2] S. Ross and J. P. Olivier, (1964) *On Physical Adsorption*, Interscience, New York, p. 8.

[3] C. Sandford and S. Ross, (1954) *J. Phys. Chem.*, **58**, 228.

[4] C. Gurney, (1949) *Proc. Phys. Soc.* (London), **A62**, 639.

[5] C. Herring, (1953), in *Structure and Properties of Solid Surfaces*, Ed. R. Gomer and C. S. Smith, University of Chicago Press, Chicago, p. 5.

[6] F. M. Fowkes, (1964) *Ind. Eng. Chem.*, **56**, 40.

[7] A. W. Adamson, (1976) *Physical Chemistry of Surfaces*, 3rd ed., Wiley, New York, p. 249.

[8] M. M. Nicholson, (1955) *Proc. Roy. Soc. (London)*, **A228**, 490.

[9] I. F. Guilliatt and N. H. Brett, (1969) *Trans. Faraday Soc.*, **65**, 3328.

[10] P. J. Anderson and A. Scholtz, (1968) *Trans. Faraday Soc.*, **64**, 2972.

[11] W. J. Dunning, (1961), in *Adhesion*, Ed. D. D. Eley, Oxford University Press, Oxford, p. 67.

[12] J. E. Lennard-Jones and B. M. Dent, (1928) *Proc. Roy. Soc. (London)*, **A121**, 247.

[13] C. Herring, (1951) in *Physics of Powder Metallurgy,* Ed. W. E. Kingston, Mc-Graw Hill, New York.

[14] E. J. W. Verwey, (1946) *Rec. trav. chim.,* **65**, 521.

[15] G. C. Benson, P. I. Freeman and E. Dempsey, (1961) *Adv. in Chem. Ser.* No. 33, 26.

[16] G. C. Benson, P. Balk and P. White, (1959) *J. Chem. Phys.,* **31**, 109.
G. C. Benson, P. I. Freeman and E. Dempsey, (1963) *J. Chem. Phys.,* **39**, 302.
G. C. Benson, and T. A. Claxton, (1968) *J. Chem. Phys.,* **48**, 1356.

[17] J. J. Burton and G. Jura, (1967) in *Fundamentals of Gas-Surface Interactions,* Eds. H. Saltzburg, J. N. Smith and M. Rogers, Academic Press, New York, p. 75.

[18] B. J. Alder, J. R. Vaisnys and G. Jura, (1959) *J. Phys. Chem. Solids,* **11**, 182.

[19] H. H. Schmidt and G. Jura, (1960) *J. Phys. Chem. Solids,* **16**, 60.

[20] J. J. Burton and G. Jura, (1966) *J. Phys. Chem. Solids,* **27**, 961.

[21] J. S. Anderson, (1964), Liversidge Lecture, *Proc. Chem. Soc.,* 77.
M. E. Straumanis and H. W. Li, (1960) *Z. Anorg. Chem.,* **306**, 143.
S. Andersson, B. Collen, U. Kuylenstierna and A. Magnéli, (1957) *Acta Chem. Scand.,* **11**, 1641.

[22] F. E. Stone, (1955), in *Chemistry of the Solid State,* Ed. W. E. Garner, Butterworths, London, p. 45.

[23] N. N. Greenwood, (1968) *Ionic Crystals, Lattice Defects and Non-Stoichiometry,* Butterworths, London, p. 148.

[24] (a) A. H. Cottrell, (1953) *Dislocations and Plastic Flow in Crystals,* Oxford University Press, Oxford.
(b) D. Hull, (1965) *Introduction to Dislocations,* Pergamon, Oxford.
(c) W. T. Read, Jr., (1953) *Dislocations in Crystals,* McGraw-Hill, London.
(d) J. P. Hirth and J. Lothe, (1968) *Theory of Dislocations,* McGraw-Hill, New York.

[25] Ref. 7, p. 264.

[26] J. M. Burgers, (1940) *Proc. Phys. Soc. (London),* **52**, 23.

[27] F. C. Frank, (1949) *Disc. Faraday Soc.,* **5**, 48.

[28] W. K. Burton and N. Cabrera, (1949) *Disc. Faraday Soc.,* **5**, 33, 40.

[29] W. K. Burton, N. Cabrera and F. C. Frank, (1949) *Nature,* **163**, 398; (1951) *Phil. Trans. Roy. Soc. (London),* **A243**, 299.

[30] L. J. Griffin, (1950) *Phil. Mag.,* **41** 196.

[31] (a) A. R. Verma and S. Amelinckx, (1951) *Nature,* **167**, 939.
(b) A. R. Verma, (1951) *Phil. Mag.,* **42**, 1005.

[32] A.J. Forty, (1951) *Phil. Mag.,* **42**, 670.

[33] I. M. Dawson and V. Vand, (1951) *Nature,* **167**, 476.

[34] F. C. Frank, (1951) *Acta Cryst.,* **4**, 497;
Phil. Mag., **42**, 1014.

[35] K. Lonsdale, (1948) *Crystals and X-rays,* Bell & Co., London.

[36] Ref. 24a, p. 188.

[37] J. J. Bikerman, (1970) *Physical Surfaces,* Academic Press, New York, p. 169.

[38] Lewis Carroll, (1970) *Through the Looking Glass and What Alice Found There,* Revised annotated ed. by M. Gardner, Penguin, London, p. 269.

[39] R. M. Goodman, H. H. Farrell and G. A. Somorjai, (1968) *J. Chem Phys.,* **48**, 1046.

[40] W. P. Ellis and R. L. Schwoebel, (1968) *Surface Sci.,* **11**, 82.

[41] American Standards Assoc, B46.1 Am. Soc. Mech. Engrs., New York (1955). See also ref. 37, p. 181.

[42] P. H. Emmett, (1942) *Adv. Colloid Sci.,* **1**, 1.

[43] J. L. Shereshefsky and C. E. Weir, (1936) *J. Amer. Chem. Soc.,* **58**, 2022.

[44] V. Veselovskii, (1939) *Acta Physicochim., URSS,* **11**, 815.

[45] T. L. O'Connor and H. H. Uhlig, (1957) *J. Phys. Chem.,* **61**, 402.

[46] D. Altenpohl, (1965) *Aluminium and Aluminiumlegierungen.* Springer-Verlag, Berlin.

[47] G. Beilby, (1921) *Aggregation and Flow in Solids,* Macmillan, New York.

[48] R. C. French, (1933). *Proc. Roy. Soc. (London),* **A140**, 637.

[49] H. Raether, (1948) *Z. Physik.,* **124**, 286.

[50] F. P. Bowden and D. Tabor, (1964) *The Friction and Lubrication of Solids,* Part II, The Clarendon Press, Oxford.

[51] A. R. Bailey, (1954) *Rep. Prog. Appl. Chem.,* **34**, 85.

[52] T. S. Renzema, (1952) *J. Appl. Phys.,* **23**, 1142.

[53] W. Cochrane, (1928) *Proc. Roy. Soc. (London),* **A166**, 228.

[54] J. J. Trillat, (1947) *Compt. Rend.,* **224**, 1102.

[55] H. P. Schreiber and G. C. Benson, (1955) *Can. J. Phys.,* **33**, 534.

[56] J. T. Davies and E. K. Rideal, (1963) *Interfacial Phenomena,* 2nd ed., Academic Press, New York, p. 5.

[57] P. V. Hobbs and B. J. Mason, (1964) *Phil. Mag.,* **9**, 181.

[58] H. H. G. Jellinek and S. H. Ibrahim, March 1966, *Abs. of 151st Meeting Amer. Chem. Soc.,* Pittsburgh.

[59] B. Chalmers, R. King and R. Shuttleworth, (1948) *Proc. Roy. Soc. (London),* **A193**, 465.

[60] G. C. Kuczynski, (1949) *J. Metals,* **1**, 129; Discussion, (1949) *J. Metals,* **1**, 897.

[61] H. Udin, A. J. Shaler and J. Wulff, (1949) *J. Metals,* **1**, 186; Discussion, (1949) *J. Metals,* **1**, 894.

[62] G. Tammann, (1925) *Z. Anorg. Chem.,* **194**, 67; (1928) **176,** 46.

[63] G. I. Finch and K. P. Sinha, (1957) *Proc. Roy. Soc. (London),* **A239**, 145.

[64] D. J. M. Bevan, J. P. Shelton and J. S. Anderson, (1948) *J. Chem. Soc.,* 1729.

[65] G. F. Huttig, (1942) *Kolloid Z.,* **98**, 263.

[66] Ref .5, p. 1.

[67] G. C. Kuczynski, (1956) *Acta Metall.*, **4**, 58.
[68] T. J. Wiseman, (1976) in *Characterisation of Powders,* Eds. G. D. Parfitt and K. S. W. Sing, Academic Press, London, p. 167.
[69] D. Barby in Ref 68, p. 403.
[70] A. I. Medalia and D. Rivin in Ref. 68, p. 304.
[71] P. Jackson and G. D. Parfitt, (1972) *J. Chem Soc. Faraday Trans. II,* **68**, 896.
[72] J. A. Hockey and P. Jones, (1971) *Trans. Faraday Soc.,* **67**, 2679.
[73] J. C. R. Waldsax and M. J. Jaycock, (1971) *Disc. Faraday Soc.,* **52**, 215.
[74] M. J. Jaycock and J. C. R. Waldsax, (1974) *J. Chem. Soc. Faraday Trans. I,* **70**, 1'501.
[75] P. C. Carman, (1950) *Trans. Faraday Soc.,* **36**, 964.
[76] W. J. de Bruin and Th. van der Plas, (1964) *Rev. Gen. Caout.,* **41**, 452.
[77] H. P. Boehm, (1966) *Adv. Catalysis,* **16**, 198.
[78] B. R. Puri, (1976) in *Chemistry and Physics of Carbon,* Ed. P. L. Walker, Jr., Marcel Dekker, New York, Vol. 6, p. 191.
[79] D. Rivin, (1963) in *Proceedings of the Fifth Conference on Carbon,* Pergamon, New York, Vol. 2, p. 199.
[80] J. E. Lennard-Jones, (1932) *Trans. Faraday Soc.,* **28**, 332.
[81] D. O. Hayward and B. M. W. Trapnell, (1964) *Chemisorption,* Butterworths, London, p. 88.
[82] Ref. 2, p. 150.
[83] E. W. Sidebottom, W. A. House and M. J. Jaycock, (1976) *J. Chem. Soc. Faraday Trans. I,* **72**, 2709.
[84] see J. H. de Boer, (1968) *The Dynamic Character of Adsorption,* 2nd ed. Oxford, p. 30.
[85] S. Kruyer, (1955) *Proc. K. ned. Akad. Wet.,* **B58**, 73.
[86] J. H. de Boer, (1956) *Adv. Catalysis,* **8**, 85, see also Ref. 84, p. 233.
[87] N. N. Avgul, A. V. Kiselev, I. A. Lygina and D. P. Poschkus, (1959) *Izv. Adad. Nauk SSR, Otd. Khim. Nauk* 1196; (1959) *Bull Acad. Sci USSR, Div. Chem. Sci.* 1155 — English translation.
[88] F. Ricca, C. Pisani and E. Garrone, (1969) *J. Chem. Phys.,* **51**, 4079, *errata* in (1970) *J. Chem. Phys.,* **53**, 2546.
[89] F. Ricca and E. Garrone, (1970) *Trans. Faraday Soc.,* **66**, 959.
[90] F. Ricca, C. Pisani and E. Garrone, (1972) *Proc. 2nd Int. Symp. Adsorption-Desorption Phenomena,* Academic Press, New York, p. 111.
[91] W. A. Steele, (1974) *The Interaction of Gases with Solid Surfaces,* Pergamon, Oxford, p. 42.
[92] W. A. House and M. J. Jaycock, (1974) *J. Chem. Soc., Faraday Trans. I,* **70**, 1348.
[93] W. A. House and M. J. Jaycock, (1975) *J. Chem. Soc. Faraday Trans. I,* **71**, 1597.
[94] E. Madelung, (1918) *Z. Physik.,* **19**, 524.

[95] M. Born, (1923) *Atomtheorie des festen Zustandes*, Leipzig, 2nd ed. p. 715.

[96] J. E. Lennard-Jones and B. M. Dent, (1928) *Trans. Faraday Soc.*, **24**, 92.

[97] T. L. Hill, (1960) *Statistical Thermodynamics*, Addison-Wesley Pub. Co. Inc., Reading, U.S.A., p. 124, 172.

[98] G. S. Rushbrook, (1949) *Introduction to Statistical Mechanics*, Oxford University Press, London, p. 74.

[99] T. Hayakawa, (1957) *Bull. Chem. Soc. Japan*, **30**, 124, 236, 243, 332, 337.

[100] Ref. 2, p. 14.

[101] Ref. 2, p.19.

[102] Ref. 84, p. 132.

[103] W. A. House and M. J. Jaycock, (1977) *J. Colloid Interface Sci.*, **59**, 252.

[104] J. H. de Boer and J. C. P. Broekhoff, (1967) *Proc. K. Ned. Akad. Wet.*, **B70**, 317.

[105] I. Langmuir, (1918) *J. Amer. Chem. Soc.*, **40**, 1361.

[106] R. H. Fowler, (1936) *Proc. Camb. Phil. Soc.*, **32**, 144.

[107] D. M. Young and A. D. Crowell, (1962) *Physical Adsorption of Gases*, Butterworths, London, p. 108.

[108] R. H. Fowler, (1935) *Proc. Camb. Phil. Soc.*, **31**, 260.

[109] R. H. Fowler and E. A. Guggenheim, (1956) *Statistical Thermodynamics*, Cambridge Univ. Press, p.426.

[110] Ref. 109, p. 430.

[111] R. A. Pierotti and H. E. Thomas (1971), in *Surface and Colloid Science*, Ed. E. Matijevic, Wiley-Interscience, New York, Vol. 4, p. 93.

[112] Ref. 2, p. 123.

[113] A. W. Adamson and I. Ling, (1961) *Adv. Chem. Ser.*, No. 33, 51.

[114] S. Ross and I. D. Morrison, (1975) *Surface Sci.*, **52**, 103.

[115] W. A. House and M. J. Jaycock, (1978) *Colloid and Polymer Sci.*, **256**, 52.

[116] L. Berenyi, (1920) *Z. Phys. Chem.*, **94**, 628.

[117] W. A. House, (1975) PhD thesis, Loughborough University of Technology.

[118] W. A. House and M. J. Jaycock, (1977) *J. Colloid Interface Sci.*, **59**, 266.

[119] W. A. House, (1978) *J. Colloid Interface Sci.*, **67**, 166.

[120] R. S. Sacher and I. D. Morrison, (1979) *J. Colloid Interface Sci.*, **70**, 153.

[121] S. Brunauer, (1945) *The Adsorption of Gases and Vapours*, Oxford Univ. Press and Princeton Univ. Press, Princeton, New Jersey, U.S.A.

[122] S. Brunauer, L. S. Deming, W. S. Deming and E. Teller, (1940) *J. Amer. Chem. Soc.*, **62**, 1723.

[123] L. H. Reyerson and A. E. Cameron, (1935) *J. Phys. Chem.*, **39**, 181.

[124] A. W. Adamson and L. Dormant, (1966) *J. Amer. Chem. Soc.*, **88**, 2055.

[125] E. C. C. Baly, (1937) *Proc. Soc. (London)*, **A160**, 465.

[126] S. Brunauer, P. H. Emmett and E. Teller, (1938) *J. Amer. Chem. Soc.*, **60**, 309, for errata see P. H. Emmett and T. W. de Witt, (1941) *Ind. Eng. Chem. (Anal.)*, **13**, 28.

[127] T. L. Hill, (1946) *J. Chem. Phys.*, **14**, 263.

[128] Ref. 84, p. 200.

[129] J. H. Singleton and G. D. Hasley, (1954) *J. Phys. Chem.*, **58**, 1011.

[130] R. B. Anderson, (1946) *J. Amer. Chem. Soc.*, **68**, 686.

[131] R. B. Anderson and W. K. Hall, (1948) *J. Amer. Chem. Soc.*, **70**, 1727.

[132] G. F. Huttig, (1948) *Monatsh. Chem.*, **78**, 177.

[133] T. L. Hill, (1950) *J. Amer. Chem. Soc.*, **72**, 5347.

[134] F. C. Tompkins, (1950) *Ann. Rep. Prog. Chem.*, **60**.

[135] G. Jura and W. D. Harkins, (1943) *J. Chem. Phys.*, **11**, 430.

[136] M. Polanyi, (1914) *Verh. Dtsch. Phys. Ges.*, **16**, 1012; (1916) **18**, 55.

[137] L. Berenyi, (1923) *Z. Phys. Chem.*, **105**, 55.

[138] Y. I. Frenkel, (1946) *Kinetic Theory of Liquids*, The Clarendon Press, Oxford. (Dover Publications reprint (1955)).

[139] G. D. Hasley, Jr., (1948) *J. Chem. Phys.*, **16**, 931.

[140] T. L. Hill, (1952) *Adv. Catal.*, **4**, 211.

[141] Ref. 107, p. 170.

[142] J. H. de Boer and C. Zwikker, (1929) *Z. Phys. Chem.*, **B3**, 407.

[143] G. D. Halsey, Jr., (1951) *J. Amer. Chem Soc.*, **73**, 2693.

[144] Ref. 7, p 582.

[145] A. G. Keenan and J. M. Holmes, (1949) *J. Phys. Colloid Chem.*, **53**, 1309.

[146] S. J. Gregg and K. S. W. Sing, (1967) *Adsorption, Surface Area and Porosity*, Academic Press, London, p. 90.

[147] J. F. Duncan, (1949) *Trans. Faraday Soc.*, **45**, 879.

[148] W. D. Harkins and G. Jura, (1944) *J. Amer. Chem. Soc.*, **66**, 1366.

[149] M. L. Corrin, (1951) *J. Amer. Chem. Soc.*, **73**, 4061.

[150] Ref. 107, Ch. 8.
Ref. 146, Ch. 8.
Ref. 2, Ch. 2.

[151] M. J. Jaycock (1978) in *Particle Size Analysis*, Ed. M. J. Groves, Heyden, London, p. 308.

[152] Ref. 2, p. 39.

[153] A. J. Rosenberg and C. S. Martel, (1957) *J. Phys. Chem.*, **61**, 512.

[154] Ref. 2, p. 34.

[155] W. A. House and M. J. Jaycock, (1975) *J. Appl. Chem. Biotech.*, **25**, 327.

[156] H. Melville and B. G. Gowenlock, (1964) *Experimental Methods in Gas Reactions*, Macmillan, London, p. 198; A. S. Friedman and D. White, (1950) *J. Amer. Chem. Soc.*, **72**, 3931.

[157] W. H. Keesom, J. Mazur and J. J. Meihuizen, (1935) *Physica*, **2**, 669.

[158] R. A. Beebe, J. B. Beckwith and J. M. Honig, (1945) *J. Amer. Chem. Soc.*, **67**, 1554.

[159] H. L. Pickering and H. C. Eckstrom, (1952) *J. Amer. Chem. Soc.*, **74**, 4775.

[160] R. T. Davis, T. W. DeWitt and P. H. Emmett, (1947) *J. Phys. Chem.*, **51**, 1232.

[161] R. A. W. Haul, (1956) *Angew. Chem.,* **68**, 238.

[162] J. H. Singleton and G. D. Halsey, (1955) *Canad. J. Chem.,* **33**, 184.

[163] L. Bonnetain, X. Duval, M. Letort and P. Souny, (1956), C. R. *Acad. Sci.,* Paris, **242**, 1979.

[164] A. Thomy and X. Duval, (1970) *J. Chim. Phys.,* **67**, 1102.

[165] Ref. 146, p. 85.

[166] P. H. Emmett and S. Brunauer, (1937) *J. Amer. Chem. Soc.,* **59**, 1553, 2682.

[167] M. L. Corrin, (1951) *J. Amer. Chem. Soc.,* **73**, 4061.

[168] E. H. Loeser and W. D. Harkins, (1950) *J. Amer. Chem. Soc.,* **72**, 3427.

[169] see S. J. Gregg and K. S. W. Sing, (1976) in *Surface and Colloid Science,* Ed. E. Matijevec, Wiley-Interscience, New York, Vol. 9, p. 231.

[170] M. M. Dubinin, (1965) *Russ. J. Phys. Chem. (Eng. Translation),* **39**, 697; (1955) *Quart. Rev.,* **9**, 101.

[171] Ref. 169, p. 318.

[172] Ref. 146, p. 225.

[173] M. M. Dubinin, (1957) *Proc. Conf. Ind. Carbon Graphite,* 1957, SCI London p. 219.

[174] M. G. Kagamer, (1959) *Zhur Fiz. Khim.,* **33**, 2202.

[175] Ref. 169, p. 323.

[176] R. Zsigmondy, (1911) *Z. Anorg. Chem.,* **71**, 356.

[177] E. O. Kraemer, (1931) in *Treatise on Physical Chemistry,* Ed. H. S. Taylor, Van Nostrand, London, p. 1663.

[178] J. W. McBain, (1935) *J. Amer. Chem. Soc.,* **57**, 699.

[179] C. Pierce, (1953) *J. Phys. Chem.,* **57**, 149.

[180] C. Orr and J. M. Dalla Valle, (1959) *Fine Particle Measurement,* Macmillan, London, p. 271.

[181] M. R. Harris and G. Whitaker, (1963) *J. Applied Chem.,* **31**, 348.

[182] B. C. Lippens and J. H. de Boer, (1965) *J. Catal.,* **4**, 319.

[183] J. H. de Boer, B. C. Lippens, B. G. Linsen, J. C. P. Broekhoff, A. van den Heuvel and Th. J. Osinga, (1966) *J. Colloid Interface Sci.,* **21**, 405.

[184] K. S. W. Sing, (1968) *Chem. Ind. (London),* 1520.

[185] K. S. W. Sing, (1970) in *Surface Area Determination,* Eds. D. H. Everett and R. H. Ottewill, Butterworths, London, p. 25.

[186] D. W. McKee, (1959) *J. Phys. Chem.,* **63**, 1256.

CHAPTER 5

The interface between a liquid and a solid

5.1 INTRODUCTION

Some of the most important and extensive applications of surface chemistry are concerned with solid-liquid interfaces, for example detergency, lubrication, wetting of powders, and precipitation. The physical aspects of this interface are manifest in the phenomena of wetting and spreading, and measurements of surface free energies and contact angles have given a clearer insight into the nature and effect of the forces that act at the interface. This chapter is concerned with the contact between a liquid and a solid, both in terms of the macroscopic effects which arise and the microscopic events at the interface with which they are associated. We shall give most consideration to pure liquids since we have a better fundamental understanding of the appropriate forces involved in molecular contact with a solid. Nevertheless the majority of the practical applications involving the solid-liquid interface are concerned with two or more components in the liquid phase, and it is therefore necessary to consider the distribution of these components at the interface and the effect of this distribution on the macroscopic phenomena.

5.2 WETTING

A drop of liquid, free in space, is drawn into a spherical shape by the tensile forces of its surface tension, and the magnitude of this force, that is, the magnitude of the surface free energy of the liquid, decreases in the order metals > hydrogen bonded compounds > polar compounds > non-polar compounds. When a drop of liquid is brought into contact with a flat solid surface, the final shape taken up by the drop depends on the relative magnitudes of the molecular forces that exist within the liquid (cohesive) and between liquid and solid (adhesive). The index of this effect is the contact angle which the liquid subtends with the solid. It is generally found that liquids with low surface tension easily wet most solid surfaces, giving a zero contact angle — the molecular adhesion between solid and liquid is greater than the cohesion between the molecules of the liquid. Liquids with high surface tension mostly give a finite contact angle, and here the cohesive forces become dominant.

The situation is illustrated in Fig. 5.1. Young originally considered (qualitatively) the equilibrium in 1805, and Dupré in 1869 put it in mathematical terms. The simplest argument is to consider the forces that exist at the three-phase contact (Fig. 5.1(b)). Let the contact region move so that an additional m^2 of solid is wetted, then there is a surface free energy increase at the solid-liquid interface (γ^{SL}), a decrease at the solid-vapour interface (γ^{SV}) and an increase at the liquid surface ($\gamma^{LV}\cos\theta$). By the principle of virtual work

$$\gamma^{SV} = \gamma^{SL} + \gamma^{LV}\cos\theta \tag{5.1}$$

which is the so-called Young (or Young-Dupré) equation and is the same as that obtained by taking the horizontal components of the surface forces. It may be argued that equation (5.1) is not correct since no account is taken of the vertical component of γ^{LV}. There is evidence for distortion of a solid surface by this component, and therefore strictly the stresses in the solid which in part balance the vertical force, should be included in the derivation of the contact angle equilibrium. However, for most practical cases where the solid is thick and the experiment is carried out at a temperature at which it is not subject to elastic and plastic deformation, the neglect of the effects is usually justified.

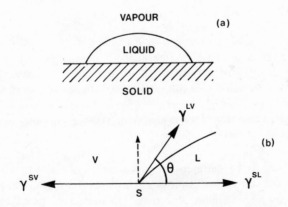

Fig. 5.1 – The shape of a drop of liquid in contact with a solid surface when $\theta < 90°$ (a), and the forces that exist at the three phase contact (b).

Experimental verification of the Young expression has been provided by Bailey and Kay [1], using measurement of the energies involved in the bifurcation of mica under different conditions in order to provide values for the γ terms. In vapour for γ^{SV}, in liquid for γ^{SL}, and for liquids which give zero contact angle ($\cos\theta = 1$) the difference between these values is equal to the surface tension of the liquid. Their data are given in Table 5.1. The agreement between the data in the last two columns is considered satisfactory justification of the use of the Young equation, at least for this system. At this point it is necessary to stress that γ^{SV} refers to the solid surface with an equilibrium adsorbed film of vapour of the liquid — it is not equal to the surface energy γ^S of the bare solid. The term film pressure Π is defined as $\gamma^S - \gamma^{SV}$, and is obtained experimentally from vapour adsorption data — it is not necessarily negligible.

Table 5.1

Parameters in the Young equation (in mN m^{-1})

	γ^{SV}	γ^{SL}	$\gamma^{SV} - \gamma^{SL}$	γ^{LV}
water	182.8	107.3	75.5	72.8
hexane	271	255	16.0	18.4

The term 'wetting' is often used loosely — for practical purposes it is usually said if $\theta > 90°$ the liquid does not wet the solid, and strictly only if $\theta = 0$ is wetting considered to occur. Young's equation does not hold if $\theta = 0$, and the imbalance of surface free energies is then defined by the spreading tension.

$$\sigma^{SLV} = \gamma^{SV} - (\gamma^{LV} + \gamma^{SL}). \tag{5.2}$$

Equation (5.2) is analogous to equation (3.114). σ^{SLV} is positive if spreading is accompanied by a decrease in free energy, that is, spontaneous. Comparing equation (5.2) with the definitions of work of adhesion and cohesion, we see that the spreading tension is the difference between the work of adhesion of liquid to solid ($W_A^{SL} = \gamma^{LV} + \gamma^{SV} - \gamma^{SL}$) and work of cohesion of the liquid ($W_C^L = 2\gamma^{LV}$) — see Fig. 3.31. Again the relative magnitudes of the molecular forces are being manifest.

The wetting of a powder is a common practical problem, and to describe this it is useful to be more specific about the kinds of wetting processes that can be involved in the formation of a solid-liquid interface. Three distinct types of wetting may be defined, and these are designated as adhesional wetting, spreading wetting and immersional wetting, according to the mechanical process taking place.

(a) Adhesional wetting. The formation of 1 m^2 of solid-liquid interface by bringing into contact 1 m^2 of plane solid surface with 1 m^2 of plane liquid surface involves work W_A given by

$$W_A^{SLV} = \gamma^{SL} - (\gamma^S + \gamma^{LV}) \tag{5.3}$$

($-W_A^{SLV}$ is the work of adhesion, that is, the work required to restore the initial condition).

(b) Spreading wetting. When a drop of liquid spreads over a solid surface, for each unit area of solid surface that disappears equivalent areas of liquid surface and solid-liquid interface are formed. The work of spreading wetting is given by

$$W_S^{SLV} = (\gamma^{SL} + \gamma^{LV}) - \gamma^{SV}. \tag{5.4}$$

Note here that γ^{SV} is used since the solid is assumed to be in equilibrium with the vapour of the liquid.

(c) Immersional wetting. If 1 m² of solid surface is immersed in a liquid the work involved is simply

$$W_W^{SLV} = \gamma^{SL} - \gamma^{SV} \tag{5.5}$$

To apply these concepts to the powder problem it is useful to consider the particles as cubes and Fig. 5.2 illustrates the process of completely wetting a cube. Taking the side length as 1 m and assuming that before wetting the solid is in equilibrium with the vapour of the liquid, the energy changes that take place are given by

(a) to (b) adhesional wetting:

$$W_A^{SLV} = \gamma^{SL} - (\gamma^{SV} + \gamma^{LV}) = -\gamma^{LV} (\cos \theta + 1) \tag{5.6}$$

(b) to (c) immersional wetting:

$$W_W^{SLV} = 4\gamma^{SL} - 4\gamma^{SV} = -4\gamma^{LV} \cos \theta \tag{5.7}$$

(c) to (d) spreading wetting:

$$W_S^{SLV} = (\gamma^{SL} + \gamma^{LV}) - \gamma^{SV} = -\gamma^{LV} (\cos \theta - 1) \tag{5.8}$$

The contact angle is introduced into equations $(5.6 - 5.8)$ by making use of the Young equation (equation (5.1)).

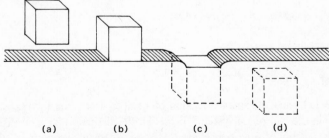

(a)   (b)   (c)   (d)

Fig. 5.2 – The three stages involved in the wetting of a solid cube: adhesional wetting (a), immersional wetting (b), and spreading wetting (c).

For practical purposes it is useful to consider under what conditions the powder would wet spontaneously. This occurs when W is negative — if positive, work must be expended on the system for the process to take place.

For the three separate stages it may be concluded that

(i) adhesional wetting is spontaneous if θ is less than $180°$,

(ii) immersional wetting is only spontaneous if θ is less than $90°$,

and (iii) spreading wetting is only spontaneous when $\theta = 0$.

For the total process

$$W_t = W_A^{SLV} + W_W^{SLV} + W_S^{SLV} = 6\gamma^{SL} - 6\gamma^{SV} = -6\gamma^{LV} \cos\theta \quad (5.9)$$

and for spontaneity θ must be less than $90°$. But since one of the separate stages requires zero contact angle, this must be the condition for spontaneous wetting for without it the powder would tend to float and some work would be required to bring about the spreading process. Surface active agents (tensides) are used to ensure θ is close to zero in the practical wetting of powders.

Several kinds of forces exist between the molecules of the two phases in contact, and following Fowkes [2] these may be expressed as a sum of work of adhesion terms for the different kinds of interactions giving the total reversible work of adhesion for the solid — liquid interface as

$$W_A^{SLV} = {}^dW_A^{SLV} + {}^hW_A^{SLV} + {}^\pi W_A^{SLV} + {}^pW_A^{SLV} + {}^eW_A^{SLV} \quad (5.10)$$

where the superscripts refer to the type of interaction; d = dispersion, h = hydrogen bond, $\pi = \pi$ − bond, p = other polar interactions, and e = contribution due to charge separation at the interface giving rise to an electric double layer. The dispersion forces are always present and ${}^dW_A^{SLV}$ is usually a dominant term. When only dispersion forces are operative, as in saturated hydrocarbons, Fowkes [3] argued that when one condensed liquid phase is brought into contact with another, the surface free energy of each liquid is reduced by the attractive force across the interface, and this force may be conveniently represented by the geometric mean of the surface energies $\sqrt{({}^d\gamma^1 {}^d\gamma^2)}$. Hence the total interfacial free energy is given by

$$\gamma^{12} = \gamma^{13} + \gamma^{23} - 2\sqrt{({}^d\gamma^1 {}^d\gamma^2)} \quad (5.11)$$

or for the solid-liquid system

$$\gamma^{SL} = \gamma^{SV} + \gamma^{LV} - 2\sqrt{({}^d\gamma^S {}^d\gamma^L)} \quad . \quad (5.12)$$

Fowkes [2] used the surface tension of a test hydrocarbon (γ^T) (as solid or liquid phase in the two phase system) to predict the dispersion force contribution to the surface energy of the other phase $({}^d\gamma^S)$ or $({}^d\gamma^L)$ using the work of adhesion obtained from

$$^{d}W_{A}^{SLV} = 2\sqrt{(\gamma^{T}\,^{d}\gamma^{L})}$$

$$^{d}W_{A}^{SLV} = 2\sqrt{(\gamma^{T}\,^{d}\gamma^{S})} \tag{5.13}$$

and $$^{d}W_{A}^{SLV} = 2\sqrt{(^{d}\gamma^{S}\,^{d}\gamma^{L})}$$

These equations define $^{d}\gamma^{S}$ and $^{d}\gamma^{L}$ and are internally consistent if the test liquid used is the same saturated hydrocarbon. For solid-liquid systems interacting by dispersion forces only, from equations (5.1) and (5.12), and including the surface pressure term Π, we have

$$\gamma^{LV}\cos\theta = -\gamma^{LV} + 2\sqrt{(^{d}\gamma^{S}\,^{d}\gamma^{L})} - \Pi \tag{5.14}$$

or

$$\cos\theta = -1 + \frac{2\sqrt{(^{d}\gamma^{S}\,^{d}\gamma^{L})}}{\gamma^{LV}} - \frac{\Pi}{\gamma^{LV}} \quad , \tag{5.15}$$

so that a plot of $\cos\theta$ versus $\sqrt{(^{d}\gamma^{L})}/\gamma^{LV}$ should be linear with a slope of a $2\sqrt{(^{d}\gamma^{S})}$. Extensive contact angle data on a variety of solids have been obtained by Zisman and coworkers [4] with homologous series of organic liquids, and their advancing angle data usually exhibit a monotonic dependence on γ^{LV}, as expressed by the empirical relation

$$\cos\theta = 1 - \text{constant } (\gamma^{LV} - {}^{c}\gamma), \tag{5.16}$$

where ${}^{c}\gamma$ is the value of γ^{LV} at $\cos\theta = 1$, that is, complete wetting. Zisman called ${}^{c}\gamma$ the 'critical surface tension', which is characteristic for the solid/liquid system, and is the minimum value of liquid surface tension for wetting of the solid — higher values of γ^{LV} give finite contact angles. Zisman's data are also linear when plotted according to equation (5.15), and the values of $^{d}\gamma^{S}$ and ${}^{c}\gamma$ are equal for saturated hydrocarbons. Representative values of $^{d}\gamma^{S}$ (mJ m^{-2}) are for polytetrafluorethylene (PTFE) 19.5, paraffin wax 25.5, titanium (IV) oxide (anatase) 92, graphite 108, iron 105, titanium (IV) oxide (rutile) 140, and alumina 252. The dispersion force contribution to the surface energy of liquids has been estimated [3] by this approach, by measurement of contact angle on solid hydrocarbons; some examples are given in Table 5.2.

The magnitude of polar interactions of liquids with solids can also be estimated from the values of $(^{d}\gamma^{S}\,^{d}\gamma^{L})$, and some examples are given in Table 5.3, heptane being used as the standard in each case. The effect of increasing polarity of the liquid is suitably demonstrated by the data for silica.

Table 5.2
Dispersion force contribution to surface energy of liquids.

	Surface tension/mN m^{-1}		
Liquid	$\gamma^{LV}(20°C)$	$^d\gamma^L$	other contributions
hexane	18.4	18.4	—
1-methyl naphthalene	36.4	36.4	—
tricresyl phosphate	40.9	39.2 ± 4	—
methylene iodide	50.8	48.5 ± 9	—
water	72.8	21.8 ± 0.7	51.0 ± 0.7
glycerol	63.4	37.0 ± 4	26.4 ± 4
formamide	58.2	39.5 ± 7	18.7 ± 7
dimethyl siloxane (silicone)	19.0	16.9 ± 0.5	—

Table 5.3
Energy of interaction of liquids with solids.

		W_A^{SLV}/mJ m^{-2}	
Solid	Liquid	Total interaction energy	Dispersion contribution
Graphite	heptane	48	48
	benzene	67	57
Anatase	heptane	49.4	49.4
	benzene	71	58
	1-propanol	81	53
	water	223	51
Silica	heptane	49.9	49.9
	benzene	69	59
	acetone	78	49
	1-propanol	91	49
	water	231	47
Barium Sulphate	heptane	49.4	49.4
	1-propanol	101	53
	water	232	51

Using $^d\gamma^S$ and $^d\gamma^L$ data Fowkes [3] demonstrated that water will spread (σ^{SLV} positive) only if the surface energy of the solid exceeds 243 mJ m^{-2}, hence not on graphite ($^d\gamma^S = 119$ mJ m^{-2}), nor on any of the metals (copper 60, lead 99, iron 105 mJ m^{-2}), although a thin oxide film on the metal surface might well raise the energy above the minimum value required. On ionic crystals the spreading of water is enhanced by polar forces and hydrogen bonds.

Contact angle measurement, therefore, represents an important tool for the study of the solid-liquid surface, and a number of techniques have been used with varying degrees of success. Perhaps the direct measurement of θ is the most popular; a drop is placed on a flat solid surface and the three-phase line is observed. Alternatively the dimensions of the top are measured and θ derived from the geometry, assuming a spherical segment. Reasonable precision ($\pm 0.3°$) is normally obtained, but agreement between laboratories is often not as good. Poor reproducibility is usually to be associated with hysteresis effects. Take a drop resting on a horizontal solid surface and add a small volume of the same liquid to the drop — the drop gets taller but its base dimension does not change, hence the contact angle increases. Similarly removal of liquid from the drop causes a reduction in contact angle. Addition of liquid gives *advancing* contact angles, and removal gives *receding* angles, and for each system there is a maximum value of the former before the three-phase line is broken and a minimum value of the latter before the drop contracts. The difference between advancing and receding angles can be quite large — for mercury on steel a difference of 154° has been recorded — and for practical use it is important to establish to which measurement reference is being made. Contact angle hysteresis is particularly well demonstrated by a rain drop on a dirty glass window. Surface roughness and surface heterogeneity have been identified as reasons for the hysteresis [5]. Surface roughness is an important factor, but this depends on the scale of the roughness compared with the size of the drop — all (except, perhaps, freshly cleaved mica) surfaces are rough on a microscopic scale and for these we might also expect impurities to play a major role. Surface preparation is also important, besides the need to define the parameters of the measurement being made, and because of these difficulties there is obviously some limitation in the extent to which such wetting phenomena can be used to define the molecular forces that exist at the solid-liquid interface.

Both hysteresis and temperature effects are demonstrated in Padday's experiments [6] on the wetting of pure paraffin wax by water; the contact angles were measured by a sessile drop height technique, and the derived σ^{SLV} and W_A^{SLV} data are given in Table 5.4. These data show that the work of adhesion of water advancing on paraffin wax increases (slightly) with temperature, but for the hydrocarbon (O) — water (W) system $^dW_A^{SLV}$ calculated from $2\sqrt{(^d\gamma^{Od}\gamma^W)}$ has a negative temperature coefficient, as does that calculated from the receding angle. Furthermore the free energy associated with the wetting is larger for the receding case. Perhaps these effects may reflect the structure of water at the interface which, as Padday points out, would be different from that at the liquid-vapour interface.

Table 5.4
Wetting of paraffin wax by water — temperature and hysteresis effects.

	γ^{W}	σ^{SLV}/mJ m^{-2}		W_{A}^{SLV}/mJ m^{-2}	
	mJ m^{-2}	advancing	receding	advancing	receding
20°C	72.8	-111	$-82.(6)$	35	63
30°C	71.2	-106	$-87.(7)$	36	55
40°C	69.6	-101	$-92.(4)$	38	47
Mean temp. coefficient	-0.16	-0.5	0.5	0.1(5)	-0.8

5.2.1 Heat of wetting

Heat is evolved when a solid is immersed in a liquid, and this is the *integral* heat of wetting (or immersion). In the experimental technique for measuring this heat the solid, usually in powder form, is kept in a vacuum and broken into the liquid under its own vapour. The quantity of heat evolved, besides being a function of the nature of the solid and liquid and the intermolecular forces involved in creating the new interface, is dependent on the purity of the liquid and the pretreatment of the solid, and without appropriate attention to these factors the measured heat could be quite misleading. For example, the heat of wetting of titanium (IV) oxide (anatase) in dry benzene is 150 mJ m^{-2} and in water 520 mJ m^{-2}. Addition of 180 ppm water to benzene raises the heat to 450 mJ m^{-2}, the water having a much higher affinity for the oxide surface than for the benzene. These figures relate to a powder from which surface adsorbed water has been removed, and for titanium (IV) oxide this requires heat treatment in vacuum at $\geqslant 150°C$. A sample evacuated at room temperature and immersed in dry benzene gives a heat of wetting of about 200 mJ m^{-2}, which of course is not the true value for the titanium (IV) oxide-benzene system. There is an abundance of heat of wetting data in the literature for a variety of solid-liquid pairs, but although for each experiment precision is high there are significant variations from one laboratory to another — just as in the case of contact angle measurement the 'microscopic' state of the system is all important, and solid-liquid interfacial phenomena are critically dependent on impurities, pretreatment of solid etc.

The enthalpy change (per unit area of solid) in the immersional process is given by

$$\Delta_{W}H = h_{SL}^{s} - h_{S}^{s} - n_{L}^{s}h_{L} \tag{5.17}$$

where h_{SL}^s and h_S^s are respectively the enthalpies of the solid-liquid and solid-vacuum interfaces, h_L is the enthalpy per mole of liquid, and n_L^s is the number of moles of liquid adsorbed at the interface, that is, in a different state from those in bulk liquid. The term $n_L^s h_L$ accounts for the loss of enthalpy in the bulk liquid due to molecules being transferred to the interface on adsorption.

The enthalpy of the solid-liquid interface h_{SL}^s is made up of two terms:

(a) the enthalpy h_a^s of the adsorbed layer of liquid molecules when in equilibrium with saturated vapour of the liquid, and

(b) the enthalpy h_L^s of formation of the interface between the adsorbed layer and bulk liquid, and this is taken as equal to the enthalpy of the liquid-vapour interface, that is, the heat evolved on destruction of unit area of liquid surface, as given by

$$h_L = \gamma^{LV} - T(\partial\gamma^{LV}/\partial T). \tag{5.18}$$

which can be derived for a closed system at constant P and T, from equations (2.7) and (3.56). Strictly this assumption means that the adsorbed layer at the interface must be sufficiently thick that the energies of the surface of the layer are not affected by the underlying solid, that is, it is a duplex film. Molecules adsorbed from the vapour phase on such a film do so with a heat evolved equal to the heat of liquefaction of the liquid. For a number of systems this occurs at about monolayer coverage of the solid surface by liquid molecules, but how much adsorbate is required to mask the effects of the solid depends on the nature of both adsorbate and adsorbent.

There is still considerable debate on the range of the force associated with the solid surface, and the number of layers of adsorbed molecules affected. For a flat surface, oxidised aluminium foil, contact angle and adsorption data [7] demonstrate that a monolayer (in vertical orientation) of a straight-chain alcohol containing three or more carbon atoms is sufficient to mask the high energy character of the solid surface. The extensive work of Derjaguin [8] has shown that the action of surface forces extends over large distances in polar liquids, and the curious phenomena associated with water in fine glass capillaries have been related to the long-range effect of the solid. Differential heats of adsorption (from vapour adsorption experiments) usually decrease steadily with increasing amount adsorbed, and where physical adsorption is involved tend to the heat of liquefaction of the adsorbate as the pressure approaches saturation. So the use of equation (5.18) requires consideration of the characteristics of the system under study e.g. one might hesitate for the case of an oxide in water but have more confidence with a hydrocarbon and a pure carbon black.

Hence we write

$$h_{SL}^s = h_a^s + \left[\gamma^{LV} - T(\partial\gamma^{LV}/\partial T)\right] \tag{5.19}$$

and from equations (5.17) and (5.19)

$$\Delta_{W}H = h_{a}^{s} + \left[\gamma^{LV} - T(\partial\gamma^{LV}/\partial T)\right] - h_{S}^{s} - n_{L}^{s}h_{L} \quad . \tag{5.20}$$

Now consider the immersion of a solid that has been equilibrated with the vapour of the liquid so that the surface now contains an adsorbed layer of an extent that depends on the pressure of the liquid to which it was exposed. If the adsorbed film is of the duplex type, that is, there is no further adsorption of liquid molecules on immersion, the difference between this enthalpy change $\Delta_{W}H_{D}$ and that for an outgassed surface is a direct measure of the difference between the enthalpies of molecules in the adsorbed and bulk liquid states, hence of the enthalpy of adsorption $\Delta_{a}H$. Thus

$$\Delta_{a}H = \Delta_{W}H - \Delta_{W}H_{D} \tag{5.21}$$

and

$$\Delta_{a}H = h_{a}^{s} + \left[\gamma^{LV} - T(\partial\gamma^{LV}/\partial T)\right] - h_{S}^{s} - n_{L}^{s}h_{L} - \Delta_{W}H_{D} \quad . \tag{5.22}$$

It is often assumed that $\Delta_{W}H_{D}$ may be replaced by h_{L}, but once again it is necessary for the adsorbed film to be sufficiently thick for it to be liquid-like in terms of its enthalpy. We may then write

$$\Delta_{a}H = \Delta_{W}H - \left[\gamma^{LV} - T(\partial\gamma^{LV}/\partial T)\right] \tag{5.23}$$

so that adsorption enthalpies are obtained from heat of immersion data combined with appropriate surface tension data, although $\Delta_{W}H_{D}$ is readily measured. Immersional heats as a function of pre-coverage are of particular interest since such data provide useful information on the heterogeneous nature of the solid surface [9].

The heat of immersion technique also provides a direct method of evaluating $^{d}\gamma^{S}$, the dispersion contribution to the surface free energy of a solid. For the immersional process we may define the free energy change (per unit area of solid) as

$$\Delta G^{SL} = \gamma^{SL} - \gamma^{SV} \tag{5.24}$$

and from equation (5.12) it follows that

$$\Delta G^{SL} = \gamma^{LV} - 2\sqrt{(^{d}\gamma^{L} \, ^{d}\gamma^{S})} \quad . \tag{5.25}$$

Since the enthalpy and free energy changes are related by

$$\Delta_{W}H = \Delta G^{SL} - T(\partial(\Delta G^{SL})/\partial T) \tag{5.26}$$

then

$$\Delta_W H = \gamma^{LV} - 2\sqrt{(^d\gamma^L \,^d\gamma^S)} - T\left[\frac{\partial(^d\gamma^L)}{\partial T} - \right.$$

$$\left. 2\sqrt{(^d\gamma^L)}\left(\frac{\partial\sqrt{(^d\gamma^S)}}{\partial T}\right) - 2\sqrt{(^d\gamma^S)}\left(\frac{\partial\sqrt{(^d\gamma^L)}}{\partial T}\right)\right] \qquad (5.27)$$

and values of $^d\gamma^S$ may be calculated from equation (5.27) — data for the liquid γ terms are well established, but little is known about the temperature coefficient of the dispersion contributions which nevertheless can be estimated from the fourth power of the density without introducing too much error. Zettlemoyer [10] used this method to calculate $^d\gamma^S$ from the interaction between the graphitised carbon black Graphon and some short chain hydrocarbons for which system only dispersion forces would be expected to operate. His data (for 25°C), given in Table 5.5, average to a value very close to that (70 mJ m^{-2}) established by other techniques.

Table 5.5
Calculation of $^d\gamma^S$ for Graphon from heat of wetting data.

	$\Delta_W H$/mJ m^{-2}	$^d\gamma^S$/mJ m^{-2}
hexane	103	65
heptane	112	68
octane	127	78.5

The experimental heat of wetting data in Table 5.5 show an increase with chain length of the straight chain alkane; data are available [11] for the series and show a two-fold increase in heat per unit surface area from hexane to hexadecane. Furthermore, for hexadecane the heat of wetting decreases with increasing temperature, and to explain these observations it has been postulated [12] that close to the freezing point of the liquid the layer of liquid adjacent to the solid surface has properties intermediate between those of the liquid and crystalline hydrocarbon. This concept is supported by measurements of the volume changes that accompany the wetting of Graphon by straight chain alkenes, and demonstrate a sharp increase in density parallel to a similar effect in heat of wetting as the temperature approaches the freezing point of the hydrocarbon [13]. It would appear that surface forces induce some structural order into the layers of hydrocarbon close to the interface with the long axes of the molecules lying parallel to one another, as in a crystal — a 'pre-freezing' phenomenon. How far this extends into the bulk liquid is not known, and would depend on the chemical nature of the solid/liquid pair, but in the Graphon/alkane case it seems to be a significant distance sufficient to make possible experimental measurements of parameters arising from the effect. A similar phenomenon occurs with the primary aliphatic alcohols but not with the pure monocarboxylic acids [14], and the reason for this difference in behaviour is not completely understood.

It is worth mentioning at this stage that there has been a great deal of controversy over whether a solid can have a long-range influence on the structure of liquid in contact with it, particularly with respect to water. Numerous experiments suggest that water in proximity to a solid surface is different from the bulk [8]. The first layer adjacent to the surface is subject to restricted molecular motion, and the effect can appear to be transmitted over many molecular diameters via hydrogen bonding. The layer of water so affected has been called 'vicinal water' [15], and the experiments which support or otherwise the presence of such a water layer have recently been reviewed by Peschel and Belouschek [16]. There is no unambiguous information on the thickness of the layer − it may extend up to 10 nm. But if such water exists it will be of vital importance to the activity of biological cells. Cells and tissues are abundant in interfaces, and these must exert a marked influence on adjacent water structure.

We have already seen that the forces involved in the interaction between liquid and solid are of several kinds (dispersion, polar, etc.), and for any particular solid the heat of immersion might be expected to depend on the chemical nature of the wetting liquid. For Graphon, however, the interaction is principally through dispersion forces and the heat of wetting for a series of polar and nonpolar molecules is constant within experimental error (Table 5.6) [17]. In contrast the heats for rutile titanium (IV) oxide are markedly dependent on the nature of the liquid, and the polarity of the latter is obviously important, and such data may be used to evaluate the field strength of the solid surface.

The net energy of adsorption of n_L^s mol m^{-2} is given, if the volume changes are negligible, by

$$\Delta_W H - \Delta_W H_D \cong n_L^s \left(U_a - U_l \right) \tag{5.28}$$

where subscripts a and l refer to adsorbed and liquid states. It is assumed that interactions between molecules in the adsorbed layer are the same as those in bulk liquid, so that only the surface/adsorbed molecule interaction is expressed in equation (5.28). We may divide the total interaction energy into three parts, being the contributions to London dispersion forces (U_d), to the polarisation of adsorbed molecules by the surface (U_α), and to the interaction between the permanent dipole in the adsorbate and the electrostatic field of the surface (U_μ).

$$U_a - U_l = U_d + U_\alpha + U_\mu \quad . \tag{5.29}$$

Now

$$U_\mu = - F\mu \tag{5.30}$$

where F is the average electrostatic field strength of the surface, and μ the dipole moment of the adsorbate. Hence, combining equations (5.28), (5.29) and (5.30)

$$\Delta_W H - \Delta_W H_D = n_L^s (U_d + U_\alpha) - n_L^s F\mu \tag{5.31}$$

and a plot of the left-hand side versus μ should be linear with a slope from which the field strength may be calculated. For a series of organic molecules only differing in the polar group, for example the butyl derivatives given in Table (5.6), it may be assumed that for a monolayer covered surface $\Delta_W H_D$ is approximately constant since this surface for all the derivatives is essentially paraffinic in nature; hence a plot of $\Delta_W H$ versus μ should give the same result. Chessick *et al* [18] have published such a plot for rutile titanium (IV) oxide using experimental heat of immersion data (Table 5.6) and these are illustrated in Fig. 5.3. From the slope an effective field strength of 820 MV m^{-1} was calculated for the rutile surface. For Graphon and Teflon there is an undetectable change in net interaction energy with dipole moment from which it may be assumed that the electrostatic field strength of these non-polar surfaces is zero.

Table 5.6

Heats of immersion of Graphon and rutile titanium (IV) oxide ($\Delta_w H$/mJ m^{-2}) at 25°C

	Rutile	*Graphon*
methanol	426	110
ethanol	397	110
1-butanol	410	114
2-butanol	415	115
butanal	556	—
1-nitropropane	664	—
1-aminobutane	330	106
1-chlorobutane	502	106
butanoic acid	506	115
hexane	135	103
octane	140	127

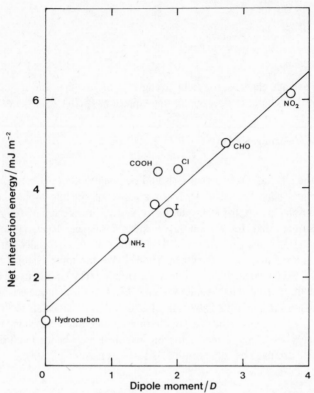

Fig. 5.3 – Heat of immersion of rutile titanium (IV) oxide in various liquids as a function of the dipole moment of the liquid phase (From Chessick *et al* [18]).

From F it is possible to calculate U_μ using equation (5.30) and U_α using

$$U_\alpha = -F^2\alpha/2 \tag{5.32}$$

where α is the polarisability of the adsorbate, giving U_d by difference. Data for rutile and Graphon are given in Table 5.7.

In a later publication Zettlemoyer [19] estimated, using Fowkes' approach, that of the 94 mJ m^{-2} calculated for U_d for 1-butanol on rutile 25 mJ m^{-2} are due to hydrogen-bonding.

The heat of immersion measurement is a potential source of fundamental data, and although high-precision calorimeters are now available commercially, the extent to which they have been used in assessing basic interfacial interactions has as yet been relatively small. Furthermore, it normally involves a high area powder, and unfortunately in most cases it is not possible to make accurate measurements of contact angle for such material. So heat of immersion and contact angle measurements are made on different materials, which may well have the same chemical composition but are subject to all the uncertainties mentioned earlier in terms of surface preparation.

Table 5.7
Interaction energies/mJ m^{-2} for Graphon and rutile titanium (IV) oxide.

	U_d	U_α	U_μ
hexane on rutile	62	30	0
hexane on Graphon	67	0	0
1-butanol on rutile	94	21	246
1-butanol on Graphon	66	0	0

5.3 ADSORPTION FROM SOLUTION AT THE SOLID-LIQUID INTERFACE

We shall now consider the interface between a solid and a liquid containing two components. At equilibrium there is a distribution of each of the components between the bulk liquid and the interphase, and in general terms this is determined by the relative strengths of the interactions of the solid surface with each component and of those between the components in the liquid phase. The composition of the interphase is a function of the concentration of solution and the temperature. There is a wealth of information on the adsorption of components from solution at the solid-liquid interface, usually expressed as adsorption isotherms; but rarely at more than one temperature, and even fewer measurements have been made of the heat effects by immersional calorimetry, hence thermodynamic data are scarce.

It is convenient to divide this subject into two parts. The first relates to solutions for which there is complete miscibility of the two components over the whole concentration range, as in binary liquid mixtures. The second is concerned with solutions that are sufficiently dilute for there to be no significant change in the activity of the solvent (the component in excess) during adsorption. This is the most common situation in practice, such as with dilute solutions of electrolytes, polymers and surface active agents which arise in many technological and natural systems. The second is essentially a special case of the first, as the following simple algebraic analysis demonstrates:

Taking m g of a solid in contact with a solution which initially contained n_1^o moles of component 1 at mole fraction x_1^o, and n_2^o moles of 2 at x_2^o, and after adsorption equilibrium is reached the solution contains n_1 moles at x_1, and n_2 at x_2, then the change on adsorption is given by

$$n_1^s m = n_1^o - n_1$$

$$= n^o x_1^o - x_1(n_1 + n_2)$$

$$= n^o x_1^o - x_1(n^o - n_1^s m - n_2^s m) \tag{5.33}$$

or

$$n^o(x_1{}^o - x_1)/m = n_1{}^s x_2 - n_2{}^s x_1 \qquad (5.34)$$

where $n_1{}^s$ and $n_2{}^s$ moles of the components are adsorbed per gram of solid and $n^o (= n_1{}^o + n_2{}^o)$ is the total number of moles of liquid components present in the system. In equation (5.34) we have the measurable quantities on the left and the unknowns on the right, and a plot of $n^o(x_1^o - x_1)/m$ versus x_1 is an adsorption isotherm but with the proviso that it is not an isotherm of actual moles of component 1 adsorbed — examination of the limiting values of the adsorption function at $x_1 = 0$ and 1 serves to illustrate the point. The measured quantity is an 'apparent' adsorption and represents the change in this quantity as the relative concentrations of the two components are varied across the whole range of mole fraction. The isotherm, being a function of the behaviour of two solution components, is commonly called a *composite* isotherm. Mathematical analysis, with assumptions, permits evaluation of the *individual* isotherms for the two components, as discussed later (p. 256).

In the case of dilute solutions, for which $x_2 \approx 1$ and $x_1 \to 0$, the measured change in solution concentration on adsorption gives directly the individual isotherm $(n_1{}^s)$ for the 'solute'.

5.3.1 Adsorption from binary liquid mixtures

The apparent adsorption is simply related to the surface excess functions which arise from the Gibbs adsorption equation with the appropriate choice of dividing surface (see section 2.7). Three cases are usually considered:

(i)   with the dividing surface located at the surface of the solid (the solid is assumed to be insoluble), the excess function being Γ_1^s;

(ii) with the dividing surface defined such that the excess of the other component is zero, giving $^2\Gamma_1$;

(iii) taking the excess of component 1 in one m^2 cross-section of surface region over the moles that would be present in a region of bulk solution containing the same total number of moles as the surface region considered, giving $^n\Gamma_1$.

The excess functions are related to each other and to the apparent adsorption (cf equation (5.34)) by the following

$$n^o \; \Delta x_1/ms = \Gamma_1^s x_2 - \Gamma_2^s x_1 = {}^n\Gamma_1 = {}^2\Gamma_1 x_2 \qquad (5.35)$$

where s is the specific surface area of the adsorbent. The reader is recommended to derive these relationships, since this helps considerably in the understanding of the principle of excess quantities.

Composite adsorption isotherms have a variety of shapes depending on the adsorption behaviour of the individual components. Ostwald and de Izaguirre [20]

were the first to point out, in 1922, that by postulating the shape of the individual isotherms the use of simple mathematical analysis led to composite
isotherms with shapes that are now well documented (Fig. 5.4). Considering the
entire range of mole fraction it is necessary for the apparent adsorption of
component 1 to go through a maximum, even if over a large part of the concentration range there is no adsorption of component 2. The apparent adsorption of
component 1 can also take negative values but only when there is adsorption of
the other component. So composite isotherms for binary liquid mixtures exhibit
one maximum (of necessity), and frequently one minimum. These principles
apply to adsorption at both liquid-vapour and solid-liquid interfaces, and there
is merit in comparing the behaviour of the two situations for any one liquid
system; this is conveniently done in terms of the excess functions $^{n}\Gamma_{1}$ and $^{n}\Gamma_{2}$.
From such a comparison it should be possible to examine the specific effect of
the solid on the adsorption process, if one exists; for example for an oxide in
contact with an alcohol + hydrocarbon mixture we might expect the adsorption
to be affected in a significant way by the alcohol-oxide interactions, in contrast
to the behaviour at the solution-vapour interface and this should be reflected in
a comparison of the composite isotherms for the two situations.

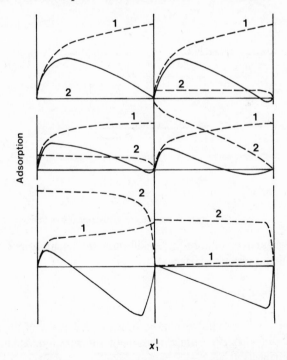

Fig. 5.4 — Shapes of composite isotherms (continuous lines) derived by simple
mathematical analysis from postulated individual isotherms for components
1 'solute', and 2 'solvent' (after Ostwald and de Izaguirre [20]).

Schay [21] has recognised five types of composite isotherms for adsorption at the solid-liquid interface, as illustrated in Fig. 5.5. These five types are also found for adsorption at the liquid-vapour interface, and an examination of the thermodynamics of this system is of value in the interpretation of isotherm shape.

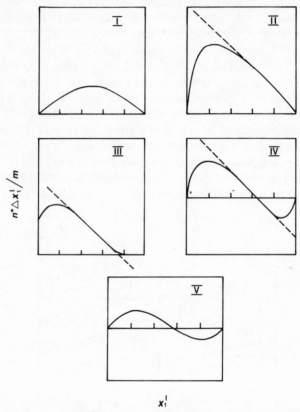

Fig. 5.5 — Schay's five types of composite isotherms (after Schay [21]).

For component i in solution at equilibrium the chemical potential in the surface phase (liquid-vapour interface) is

$$\mu_i^s = \mu_i^l + A_i\gamma^l \tag{5.36}$$

where superscript l refers to bulk solution, A_i is the molar area of the component in the surface and γ^l is the surface tension of the solution. Taking the pure liquid as the standard state

$$\mu_i^s = \mu_i^{\ominus s} + RT \ln a_i^s \tag{5.37}$$

$$\mu_i^l = \mu_i^{\ominus l} + RT \ln a_i^l \tag{5.38}$$

where a_i is the activity of the component. Hence at equilibrium

$$\mu_i^{\ominus s} + RT \ln a_i^s = \mu_i^{\ominus l} + RT \ln a_i^l + A_i \gamma^l \quad . \tag{5.39}$$

Since

$$\mu_i^{\ominus s} - \mu_i^{\ominus l} = A_i \gamma_i^l \tag{5.40}$$

then for component 1

$$RT \ln \frac{a_1^l}{a_1^s} = A_1 (\gamma_1^{LV} - \gamma^l) \tag{5.41}$$

and similarly for component 2. Hence

$$\left(\frac{a_1^s}{a_1^l} \right)^{1/A_1} \left(\frac{a_2^l}{a_2^s} \right)^{1/A_2} = K = \exp \left[(\gamma_2^{LV} - \gamma_1^{LV})/RT \right] \tag{5.42}$$

or for $A_1 = A_2 = A$

$$K = \frac{a_1^s a_2^l}{a_1^l a_2^s} = \exp \left[\frac{(\gamma_2^{LV} - \gamma_1^{LV})A}{RT} \right] \tag{5.43}$$

where K is the thermodynamic equilibrium constant for the exchange (adsorption) process as represented by the expression

$$\frac{1}{A_1} (1)^l + \frac{1}{A_2} (2)^s \rightleftharpoons \frac{1}{A_1} (1)^s + \frac{1}{A_2} (2)^l \quad . \tag{5.44}$$

From equation (5.43) it is not immediately apparent which factors predominate in deciding the shape of the composite isotherm. Interpretation of adsorption data requires accurate values of surface tension and activity coefficients, and to have such comprehensive information on any one system is not common. Some guiding principles are established. Clearly the larger the difference in surface tensions of the two pure components the larger is K, and the larger the adsorption of one component. Type I adsorption in Schay's classification would, at the liquid-vapour interface, be associated with mixtures of liquids that are similar in type, and exhibit only small deviations from ideality, for example toluene/benzene with $\gamma_2^{LV} - \gamma_1^{LV} < 1$ mN m^{-1}. For non-ideal systems with a fairly large $(\gamma_2^{LV} - \gamma_1^{LV})$, type II adsorption is obtained, for example ethanol $(\gamma^{LV} = 22.0 \text{ mN m}^{-1})$ and benzene $(\gamma^{LV} = 28.2 \text{ mN m}^{-1})$ mixture. Larger values of $(\gamma_2^{LV} - \gamma_1^{LV})$ and large departures from ideality lead to type III adsorption, and with smaller surface tension differences but with considerable non-ideality so the S-shaped curves of types IV and V adsorption result. Examples are: type III, acetone/water $(\gamma_2^{LV} - \gamma_1^{LV} = 49.69 \text{ mN m}^{-1})$; type IV, methanol/benzene $(\gamma_2^{LV} - \gamma_1^{LV} = 6.1 \text{ mN m}^{-1})$; type V, ethanoic acid/benzene $(\gamma_2^{LV} - \gamma_1^{LV} = 1.32 \text{ mN m}^{-1})$.

The interest lies in a comparison of the liquid-vapour isotherms with those for the solid-liquid interface — any differences might be interpreted in terms of specific effect of the solid surface on the composition and character of the adsorbed layer. Normally at the liquid-vapour interface it is the component with the lower surface tension that is adsorbed preferentially over the whole range of mole fraction (Types I, II and III), but where a maximum or minimum occurs in the surface tension curve an S-shaped composite isotherm is obtained (types IV and V) corresponding to an increased contribution of the other components; and it may be preferentially adsorbed over the whole range in the extreme case. At the solid-liquid interface preference may be for either component depending, it would seem, on the chemical nature of adsorbate and adsorbent, hence on the nature of the forces existing between them. Simple comparisons are made in Fig. 5.6 for chloroform + benzene mixtures with and without boehmite, and for cyclohexane + benzene mixtures at the liquid-vapour interface and on the carbon black Spheron 6, charcoal and boehmite, In the first case the component with

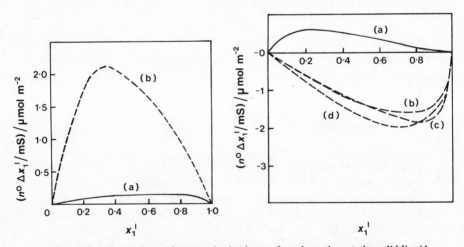

Fig. 5.6 — Comparison of composite isotherms for adsorption at the solid-liquid and liquid-vapour interfaces. Left — adsorption from mixtures of benzene and chloroform at the liquid-vapour interface (a), and on boehmite (b). Right — adsorption from mixtures of benzene and cyclohexane at the liquid-vapour interface (a), on Spheron 6 (b), on charcoal (c) and on boehmite (d) (From Kipling [25], p. 196).

the lower surface tension (chloroform) is preferentially adsorbed at both interfaces, but the adsorption is larger when the solid is present, presumably because of the enhanced interaction between the chloroform dipole and the oxide surface. In the second case that with the lower tension (cyclohexane) is preferentially adsorbed at the liquid-vapour interface, but vice versa at the solid-liquid interface reflecting the force between benzene and carbon (through π bonds) which, in a system where all the forces are relatively weak, easily tips the balance in favour of the higher surface tension component being preferentially adsorbed. Fig. 5.7 shows the simple case of 1,4-dimethylbenzene + heptane at the liquid-vapour interface, and at rutile-solution interfaces with the surface forces of the solid being modified by pre-adsorbed water [22]. With the water-free surface the S-shaped isotherm arises from the fact that the two solution components interact in a different manner with different regions of the surface (on a molecular scale), which disappears when the surface is covered with water and on which the aromatic molecule has preference. Few such analyses have been carried out since appropriate data are scarce.

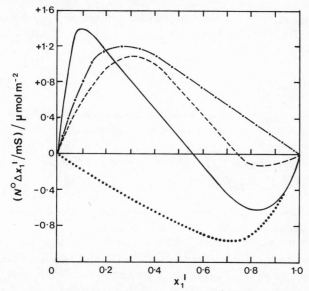

Fig. 5.7 — Composite isotherms for adsorption from 1,4-dimethylbenzene/heptane mixtures on rutile titanium (IV) oxide, free of adsorbed water (———), with approximately 1 monolayer of adsorbed water (— — — —), with 2-3 monolayers of adsorbed water (—·—·—), compared with that for the liquid-vapour interface (· · ·) (From Day *et al* [22]).

Everett's thermodynamic analysis of adsorption from binary liquid mixtures [23] demonstrates, using a lattice model and $A_1 = A_2$, that the perfect system (all activity coefficients equal to unity) gives types I and II composite isotherms, with the trend from I to II increasing with K. S-shaped isotherms are predicted by applying regular solution theory [24], although azeotropic behaviour is only evident when K is close to unity, that is, $\gamma_1^{LV} \approx \gamma_2^{LV}$, in general agreement with experiment (see above). So for non-ideal liquid mixtures the shape of the isotherm is determined by (a) the difference in adsorption energies of the two components and (b) the nature of the bulk and adsorbed phases as reflected in the activity coefficients — if (a) is large we expect types I and II, if small types IV and V with the relative proportions of positive and negative regions determined by the departure from ideality in the bulk and/or adsorbed phases. U-shaped isotherms are also predicted for molecules of different sizes assuming athermal solution behaviour, the preferential adsorption being enhanced by disparity of molecular size; for non-athermal solutions S-shaped isotherms are predicted.

In the majority of cases the composite adsorption behaviour is only of academic interest, since it gives no direct information on the actual composition of the adsorbed phase which is often required in practical application of solution adsorption. For want of anything better it has been practice to assume that the adsorption is confined to a monomolecular layer. This is acceptable to a good approximation if (a) the two molecules involved occupy approximately equal areas at the surface and (b) the adsorbate and adsorbent are such that interaction between them extends only over one molecular distance (as discussed earlier). On this basis we may take

$$n_1^s/(n_1^s)_m + n_2^s/(n_2^s)_m = 1 \qquad (5.45)$$

where $(n_1^s)_m$ and $(n_2^s)_m$ are the values of the surface concentrations when the surface is covered with a monolayer of the single component. Combining equations (5.45) and (5.34) provides sufficient data from the calculation of the individual isotherms. Monolayer values are derived from vapour adsorption data or molecular models, and although in principle they may be inaccurate for the use to which they are applied, since for the mixed monolayer the packing and orientation may differ from those in a single component monolayer, there has been a large number of such calculations made with varied success [25].

The analysis is fraught with difficulties. Take the case of a mixture of 1-octanol and 1,4-dimethylbenzene in contact with a rutile titanium (IV) oxide surface. A reasonable value of $(n_1^s)_m$ for the alcohol is equivalent to 0.2 nm^2 molec^{-1}, the cross-sectional area of a hydrocarbon chain close packed on a water surface (Langmuir-Adam trough studies; see section 3.11.4) — this assumes that all concentrations the alcohol adopts a vertical orientation. For 1,4-dimethylbenzene we take 0.45 nm^2 molec^{-1} which is obtained from a molecular model. Using these monolayer values with composite adsorption data the individual isotherms were theoretically unacceptable since they exhibited maxima and minima [26]. But for this system the chain length of the adsorbed alcohol is significantly larger than the thickness of the hydrocarbon adsorbed in parallel orientation. Hence there is effectively more than one layer of 1,4-

dimethylbenzene molecules involved in the monolayer of thickness corresponding to the length of the octanol molecule. Allowing for this effect leads to acceptable individual isotherms showing alcohol adsorption increasing with concentration in a manner (Langmuirian) that 'experience' would predict; there is no other way to predict isotherm shape. So we have 'constructed' a simple multilayer.

Everett [27] has also considered a multilayer model for monomer-dimer mixtures involving both parallel and perpendicular orientations, and shows that much larger surface excesses are to be expected than for the case when all molecules are in parallel orientation as a monomolecular layer, when there are strong positive deviations from ideality in solution, that is, the more non-ideal the greater the contribution from the second and higher layers. Multilayer theory also predicts S-shaped composite isotherms. The predominant orientations is seen to be dependent on the relative strengths of the interaction of the two segments of the dimer with the surface — if only one is preferentially adsorbed (as expected in the octanol/dimethylbenzene/rutile case considered earlier) the orientation is expected to change from parallel to perpendicular as the solution concentration is increased, although the predominant orientation is perpendicular. The concept of changing orientation with concentration makes use of equation (5.45) pointless.

To some extent theory leads practice in this field, but the experimental data required are now defined and only a few presently available satisfy all the criteria of extent and accuracy; these are summarised by Everett [28].

5.3.2 Adsorption from dilute solutions

This is the case when the solute concentration is so low that any change in solvent concentration as a result of adsorption is negligible in comparison with that of the solute — the measured concentration change in solution then gives directly the individual isotherm for the solute, that is, the value of n_1^s in equation (5.34). It must be remembered that solvent is adsorbed, but the amount involved does not enter into the simple mathematics.

Examples of adsorption at the solid-liquid interface of solutes from dilute solution are numerous, and perhaps the most useful are those for tensides (wetting agents and detergents) and for polymers, since these have wide application. It is convenient to divide the discussion into two parts — non-electrolytes and electrolytes — although there are features, for example isotherm shape, that are common to both.

5.3.2.1 *Adsorption of non-electrolytes*

A common feature of isotherms for adsorption from solution is their 'Langmuirian' shape. This term is a convenient expression of the fact that with concentration the adsorption increases with decreasing rate (that is, the isotherm is concave to the concentration axis) until a saturation level is reached. Although the mechanism of the adsorption process usually bears no resemblance to the criteria on which the Langmuir theory of adsorption is based (see page 190) — uniform energy of adsorption with coverage, etc. − the algebra is readily shown to lead to an equation of the same form as that of Langmuir. Take the adsorption process for the two component system as

$$(1)^l + (2)^s \rightleftharpoons (1)^s + (2)^l \tag{5.46}$$

assuming equal areas for the two components c.f. equation (5.44). Taking the activity coefficients of the adsorbed molecules as unity, the equilibrium constant for the process is

$$K = x_1^s \, a_2^l / x_2^s a_1^l \tag{5.47}$$

where a_1^l and a_2^l are the solvent and solute activities in solution. For a dilute solution a_1^l is constant, and writing $1/b = Ka_1^l$ we have, assuming $a_2^l = c_2^l$,

$$x_2^s = bc_2^l/(1 + bc_2^l) \tag{5.48}$$

where c_2^l is the solute concentration. Taking the number of adsorption sites on the surface as n^s mol g^{-1}, then the fraction of the surface covered by the two components is $n_2^s/n^s = \theta$ and $n_1^s/n^s = 1-\theta$ for solute and solvent respectively. Substituting $n_2^s = x_2^s n^s$ into equation (5.48) gives

$$n_2^s = n^s \, bc_2^l/(1 + bc_2^l) \tag{5.49}$$

or $\qquad \theta = bc_2^l/(1 + bc_2^l) \tag{5.50}$

which is analogous to the Langmuir expression (equation (4.77)) for adsorption of gases on solids, with c_2^l replacing the gas pressure term. The constant b is by virtue of its definition related to the energetics of adsorption. Equation (5.49) may be put in the form

$$\frac{c_2^l}{n_2^s} = \frac{1}{n^s b} + \frac{c_2^l}{n^s} \tag{5.51}$$

and a plot of c_2^l/n_2^s versus c_2^l is linear and from it values of n^s and b may be derived. Because adsorption from solution data fit expressions of the Langmuir type it does not necessarily mean that the constants (n^s and b) derived from plots of equation (5.51) have any physical significance in Langmuir terms; it may be just a question of fitting the curve to a suitable two-constant equation − care is necessary in interpretation. The reader is referred to Kipling's book [25] for more extensive debate on this subject, with examples.

Extending the argument one stage further, if we assume that the surface is heterogeneous, inferring that b is not constant but varies with coverage θ, it can be shown [29] that

$$\theta = \alpha \, (c_2^l)^{1/n} \tag{5.52}$$

where α and n are constants for the system. Equation (5.52) corresponds to an exponential distribution of adsorption heats and is the Freundlich adsorption equation. It does not become linear at low concentrations or have a saturation value, differing in these respects from the Langmuir equation. Once again, because data fit this equation it does not necessarily mean that the surface is heterogeneous. The Freundlich equation is essentially empirical, and perhaps only useful for fitting data.

Adsorption studies have been made with a variety of organic solutes from aqueous and non-aqueous solutions on to solid adsorbents ranging from the most hydrophilic oxides to the most hydrophobic graphitised carbon blacks. Numerous examples are given in Kipling's book [25]. An important variable is the pretreatment of the surface before exposure to solution, and its poor definition often contributes to lack of reproducibility (as in all surface chemistry where solids are involved).

The solubility of the solute is another major factor in controlling adsorption equilibria. In general one might anticipate an inverse relationship between the extent of adsorption of a solute and its solubility — the stronger the solute-solvent bond the smaller the extent of adsorption. This is the Lundelius rule, and it often applies. Regularity in adsorption behaviour for homologous series was originally noted by Traube who found that the surface activity (at the liquid-vapour interface) of aqueous solutions of organic solutes increases strongly and regularly as the series is ascended — this effect is known as Traube's rule. A similar effect has been observed at the solid-liquid interface, for example the adsorption of fatty acids on charcoal from aqueous solution increases in the order formic, ethanoic, propionic and butyric. The reverse of this sequence is found for adsorption on silica gel from toluene. But the reversal is also observed for adsorption from aqueous solutions on silica gel and on charcoal prepared under strongly oxidising conditions. The situation is obviously not simple. Competition for the surface between solute and solvent coupled with solute/solvent inter-actions in solution are major factors in determining which direction the adsorption will take as the series is ascended. Unfortunately only few data are available, and most are concerned with porous solids — the behaviour might well be related to the penetration of molecules into pores and the dimension of the pores. It is clear that Traube's rule is *not* a useful working rule for adsorption of dilute solutions of non-electrolytes at the solid-liquid interface.

In the adsorption of polymers it is the molecular weight and chemical structure that dominate the behaviour. Usually with simple molecular systems adsorption equilibrium is established in a short time (minutes), but for polymers this may take very much longer (days or months). So for polydisperse (broad molecular weight distribution) polymers, as most are, the smaller members are adsorbed preferentially at first and then replaced by the higher molecular weight species, as balance is established between affinity for solvent and configurational restrictions on adsorption; the lower molecular weight material is more readily desorbed (ease determined by number of segments of the molecule attached — the chance of their removal at any one time decreases the more that are attached). There is usually no difficulty in obtaining an adsorption isotherm for a polymer, and the Langmuirian shape is quite common. But interpretation in terms of the state of the polymer at the surface is not easy, and adsorption data alone are not normally sufficient to completely define the conformation. Many structures are possible for a looped or coiled polymer molecule attached in part to a surface, ranging from those which form a relatively flat and compressed layer to those giving adsorbed layers highly extended away from the surface. It is possible to estimate the fraction of polymer segments attached directly to the surface from application of infra-red spectroscopy since on adsorption there is a shift in the frequency of the characteristic band of the adsorbing group. Ellipsometry [30] provides a method for estimating the extension of loops away from the surface — by measuring the change in state of polarisation of light upon reflection from a film covered surface the average thickness of the film may be estimated. Adsorption, ellipsometric and infra-red spectroscopic data together could yield sufficient information to define the polymer conformation in the adsorbed state, but such composite data are rarely available; and there are experimental and interpretational difficulties. One such comprehensive study has recently been reported [31].

One of the most useful attempts to predict the adsorption behaviour of polymer molecules at the solid-liquid interface has been made by Clayfield and Lumb employing Monte Carlo methods [32]. Using a cubic lattice model with a 90° bond angle they generated (by computer) random polymer chains of up to 300 links with one end adsorbed on the surface, and calculated such factors as the distance of the farthest extremity from the surface and the widths of projected area occupied by the molecule. The derived configurations for the terminally adsorbed molecule indicate that the surface has little influence — there is no significant flattening of the molecule in the adsorbed state and no significant increase in the density of segments near the surface. Other Monte Carlo calculations on the configurational behaviour of adsorbed polymer molecules have been reported by Lal and Stepto [33].

Polymer adsorption is markedly dependent on the solvent; usually it is strongest from a 'poor' solvent and vice versa. Sometimes the adsorption increases with temperature, that is, it is entropy rather than energy controlled, suggesting solvent release on adsorption. These and the other facets of the adsorption of polymers have been adequately reviewed [34] and will not be considered further here.

5.3.2.2 Adsorption of electrolytes

For this discussion it is convenient to divide electrolyte adsorption into two areas, although there are many practical situations in which both are involved at the same time. The first type of adsorption involves a simple electrolyte solution in contact with a surface that carries a natural charge. This charge influences the distribution of ions in solution such that those of opposite sign to that of the surface are at a greater concentration in the vicinity of the surface, hence positively adsorbed, while the other ions are repelled from the surface and therefore negatively adsorbed. The overall effect preserves electrical neutrality, and forms what is known as the 'electric double layer' at the interface. The names of Gouy and Chapman are usually associated with this concept (see page 77), and adequate discussion of the mathematical treatment is given in standard textbooks [35]. Nevertheless, the simple picture of the 'diffuse' double layer has proved inadequate to explain various electrochemical phenomena, and the treatment was extended by Stern to include 'specific' adsorption of ions at the phase boundary (see page 80). Here there are stronger forces involved that hold the ions, probably partly solvated, at the surface, for long enough to be considered a part of the solid unit and to move with it under Brownian motion or in an electric field; these ions are found in the so called 'Stern plane'. Adsorption of counter-ions in this plane is given by

$$\frac{N_s}{n_\delta} = 1 + \frac{n_0}{n} \, \exp\left(\frac{ze\phi(\delta) + \Phi}{kT}\right) \tag{5.53}$$

assuming co-ions are not adsorbed. Here N_s is the total number of adsorption sites per unit area of surface, n_δ the number of adsorbed counter-ions per unit area, n_0 the number of solvent molecules per unit volume of solution, n the number of counter-ions per unit volume, $\phi(\delta)$ the potential at the Stern plane, and Φ the specific adsorption potential. Equation (5.53) is known as the Stern adsorption isotherm. Adsorption of simple ions into the Stern layer is relevant to the effect of ionic strength on the isoelectric point of solids derived from electrophoretic measurements, to electrocapillary phenomena, to the influence of electrolytes on the behaviour of charged monolayers at the liquid-vapour interface, to coagulation phenomena and to other colloidal processes − for further detail see reviews by Smith [36] and Adamson [37].

The second area of electrolyte adsorption concerns ionic tensides, which have great importance in detergency, mineral flotation, and a large number of industrial applications involving the formation of dispersions/suspensions of powders in liquids. The agents are of two types, anionic and cationic; examples are sodium dodecylbenzene sulphonate and cetyl trimethylammonium bromide respectively, the first being a common basic ingredient for household detergents and the second for antiseptics. Aqueous solutions of these materials behave as strong electrolytes (100% dissociation into ions in solution), and the long chain ions adsorb at the solid-solution interface to give a variety of isotherm types depending on the nature of the solid and the electrolyte, the ionic strength, the pH and the temperature. There are numerous examples in the literature, and the types have been reviewed in a qualitative manner by Giles [38]. An example of a comparatively simple system [39] is given in Fig. 5.8. It involves the anionic tenside sodium dodecyl sulphate (SDS), and the graphitised carbon black Graphon, which has near uniform graphitic-like surface. Graphon in water has a point of zero charge of pH ~ 2, so that in solutions of SDS the particles of the carbon are charged negatively. Despite this charge the long chain ions are adsorbed at the interface, a result of the tendency of hydrocarbon chains to escape from water. At low concentrations the ion adsorbed exhibits apparent Langmuirian behaviour tending to a plateau at about 4×10^{-3} mol dm^{-3}, after which the adsorption again increases until the saturation limit is reached at the critical micelle concentration (CMC) of 8×10^{-3} mol dm^{-3}. Two orientations of the adsorbed long chain ion are suggested, at low concentration parallel to the surface (area occupied at the point of inflection 0.70 nm^2) hence reducing the hydrocarbon-water interface as much as possible, and at the higher concentration a preferred vertical orientation (head groups outside), although because of repulsion not sufficiently close packed to remove all the hydrocarbon-water interface. The adsorption process has similar characteristics to that of micelle formation, so it is not surprising that the adsorbed layer at saturation has a marked similarity to the structure of a micelle, and is at equilibrium with micelles etc. in solution. Addition of neutral electrolyte (NaCl) depresses the CMC, and the concentration at which saturation adsorption is reached. Also the adsorption is increased owing to the screening

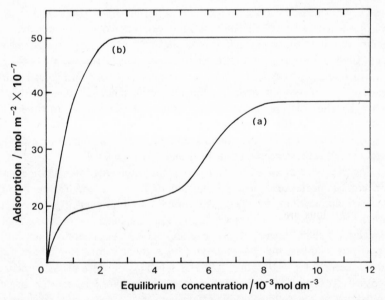

Fig. 5.8 – Adsorption of sodium dodecyl sulphate on to Graphon from aqueous solutions (a) and from solutions in 0.1 mol dm^{-3} sodium chloride (b) (From Greenwood *et al* [39]).

effect of the added electrolyte on the charged head groups, and at 0.1 mol dm^{-3} NaCl the area occupied per dodecyl sulphate ion at the surface is 0.33 nm^2. The repulsion effect has been significantly reduced, allowing the chains to pack quite close together at the interface.

With oxides the pH is perhaps more relevant since by its variation the surface charge can be varied between quite wide limits of magnitude and sign. Fuerstenau and coworkers have carried out a comprehensive study of ionic surfactant adsorption on oxide surfaces also with reference to chain length, nature of head group etc. in connection with mineral flotation studies. A review of this work is worthy of study [40].

5.3.3 Thermodynamic data from heats of immersion

The analysis of the various enthalpy terms involved when a solid is immersed in a pure liquid is readily extended to solutions by taking account of the fact that the various components of the solution may be present in the adsorbed layer, and that on adsorption a concentration change takes place in solution.

Consider a two-component solution containing initially n_1 and n_2 moles of components 1 and 2 respectively, and after immersion there are n_1^s and n_2^s moles of these components in unit area of the adsorbed layer. By simple addition of the enthalpies of the two solution components and of the outgassed solid before and after immersion, the total change in enthalpy $\Delta_W H_S$ during the immersional process is given by

$$\Delta_W H_S = h_{S1}^s - h_S^s - n_1^s \bar{h}_1^l - n_2^s \bar{h}_2^l - \ n_1(\bar{h}_1^{l*} - \bar{h}_1^l) + n_2(\bar{h}_2^{l*} - \bar{h}_2^l) \quad (5.54)$$

where h_{S1}^s is the enthalpy of the *solid-solution* interface, and $\bar{h}_1^{l*}$ the partial molar enthalpy of molecules 1 in the bulk solution before adsorption and $\bar{h}_1^{l*}$ after adsorption. The terms outside the square brackets are analogous to those of the pure liquid, as given in equation (5.17), while those within the square brackets represent the enthalpy change in bulk solution due to the change in concentration.

For a dilute solution the terms in the square brackets may be neglected, and the heat of immersion is then given by

$$\Delta_W H_S = h_{S1}^s - h_S^s - n_1^s \bar{h}_1^l - n_2^s \bar{h}_2^l \quad (5.55)$$

and the differential heat of adsorption of component 1 is

$$\Delta \bar{H}_1 = \bar{h}_1^s - \bar{h}_1^l = \left(\frac{\partial \Delta_W H_S}{\partial n_1^s} \right)_{P, T, n_2^s} \quad (5.56)$$

which can be derived from immersional heats and an adsorption isotherm.

An alternative approach [41] considers that as for pure liquids, h^s_{S1} is made up of two parts, that of the adsorbed layer h^s_a and that for interface formation h^s_L. It is assumed that the molecular interactions in bulk solution are the same as those in the adsorbed layer, and need not be included in the total enthalpy change. Taking the fraction of surface covered by component 1 as θ then the enthalpy change for this component is (a) for adsorption $\theta(h^s_{S1} - h^s_S)$ and (b) for interface formation $\theta(h^s_{1l} - h^s_{Sl})$, where h^s_{S1} is the enthalpy of surface covered with molecules of component 1, and h^s_{1l} is the enthalpy of the adsorbed layer (of 1) in contact with solution. Similarly for component 2 we have $(1-\theta)$ $(h^s_{S2} - h^s_S)$ and $(1-\theta)$ $(h^s_{2l} - h^s_{S2})$. The total change on immersion is given by

$$\Delta_W H_S = \theta(h^s_{S1} - h^s_S) + (1-\theta)\,(h^s_{S2} - h^s_S) + \theta(h^s_{1s} - h^s_{S1}) + (1-\theta)\,(h^s_{2s} - h^s_{S2})\ .$$

$$(5.57)$$

Equation (5.57) is analogous to (5.54) except the solution terms have been excluded. Young, Chessick and Healey [41] verified the validity of equation (5.57) using the Graphon 1-butanol/water system for which adsorption of 1-butanol reaches saturation at low concentration. Data for the individual terms on the right-hand side were obtained from adsorption experiments and from the immersional heats for the individual components, and allowance was made for the enthalpy of dilution which they determined from heat of solution experiments. The agreement between the measured heats of immersion for the system and those calculated using equation (5.57) is excellent, and suggests that there is no appreciable contribution to the heat of immersion from interaction between molecules in the adsorbed layer, that is, the adsorption can be treated as the sum of independent effects.

When more concentrated solutions are considered, and particularly for the whole mole fraction range of two miscible liquids, any assumptions that lead to the elimination of solution enthalpy terms would, other than in exceptional cases, appear to be invalid. Everett [23] has analysed the case for a perfect liquid mixture, and assuming monolayer adsorption shows that the heat of immersion in the mixture is related to those for the individual components by

$$\Delta_W H_S = x^s_1\,\Delta_W H_1 + x^s_2\,\Delta_W H_2 \qquad\qquad (5.58)$$

where x^s_1 and x^s_2 are the mole fractions in the adsorbed layer. Data on the charcoal/benzene/cyclohexane system show good agreement with equation (5.58) over the entire concentration range [42]. Extra terms must be included for solutions not exhibiting perfect behaviour but such terms are difficult to evaluate.

Furthermore the validity of assuming a monolayer is subject to debate as been said earlier, and some more recent heat of immersion data [43] on the apparently simple system Graphon/heptane/hexadecane strongly suggest that more than one layer of adsorbed molecules are being affected by the underlying surface. Heats of immersion were measured over the whole mole fraction range, at temperatures between $10°$ and $45°C$, and are shown in Fig. 5.9. From a comparison of these results with adsorption from solution data it was concluded that the plateaux correspond to the completion of a monolayer of hexadecane molecules oriented parallel to the solid surface. The excess heats at higher solution concentrations are attributed to the formation of a more structured phase in the adjacent surface layers, since as the temperature is lowered the excess heat increases at any one solution concentration. This is the same concept as discussed earlier as a 'pre-freezing' phenomenon observed with straight chain alkanes and Graphon. The effect shows up clearly in the heat of immersion experiment, but is not recognised from adsorption data, indicating that in the surface layers affected the solution concentration is very close to that of the bulk liquid phase.

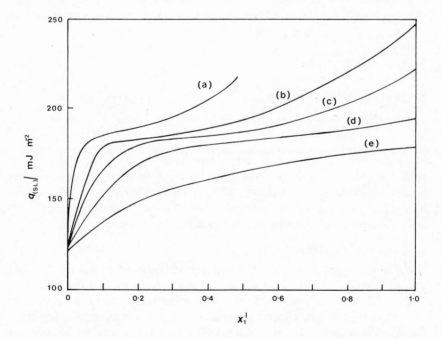

Fig. 5.9 – Heats of immersion for Graphon in heptane/hexadecane solutions at $10°C$ (a), $20°C$ (b), $25°C$ (c), $35°C$ (d) and $45°C$ (e) (From Parfitt and Tideswell [43]).

5.4 PRACTICAL APPLICATIONS

Several important applications of phenomena occurring at the solid-liquid interface will now be considered although, of necessity, in brief. The relevant principle is stressed and the reader is left to investigate the extensive literature that exists on each topic.

5.4.1 Capillarity

The phenomenon of capillary rise is well known (equation (3.24)). The potential energy that is gained by the liquid is equivalent to raising a mass $\pi r^2 hp$ of liquid through a height $h/2$. Thus the gain in potential energy is given by

$$U_1 = \tfrac{1}{2}\pi r^2 h^2 \rho g. \tag{5.59}$$

With the liquid and solid in contact at rest, and assuming zero contact angle, from equation (5.1)

$$\gamma^{SV} = \gamma^{LV} + \gamma^{SL}. \tag{5.60}$$

When the liquid moves along the tube so as to cover one m^2 of surface, the energy liberated is $\gamma^{SV} - \gamma^{SL}$ (there is no change in the area of liquid surface). Hence γ^{LV} is released for each m^2 wetted, and for height h the energy is

$$U_2 = \gamma^{LV} 2\pi r h \tag{5.61}$$

and substituting from equation (3.24)

$$U_2 = \pi r^2 h^2 \rho g, \tag{5.62}$$

which is twice the gain in energy U_1. Thus the energy liberated in the wetting process is used to pull the liquid up the tube. Equation (5.60) implies that there is sufficient available to raise it to a height $2\,h$, and if the liquid were non-viscous it would rise to that height and oscillate between 0 and $2\,h$ with its mean position at h − in practice the viscosity rapidly consumes the excess energy.

5.4.2 Wetting of textiles and powders

The rise of a liquid in a cylindrical tube is readily described, and the principle can be extended to more complex practical situations such as the wetting of textiles and powders. The spaces between the fibres in textiles and the particles in powders may as a first approximation be considered as an array of capillaries through which the liquid penetrates, and a convenient way to describe the process is in terms of the pressure P that would be required to force the liquid into the capillaries, alternatively to restrain its entry. From equation (3.10) this is given by

$$P = -2\gamma^{LV} \cos\theta/r \tag{5.63}$$

and therefore the flow is only spontaneous (P negative) when the contact angle $\theta < 90°$. If θ is not zero then using equation (5.1)

$$P = -2(\gamma^{SV} - \gamma^{SL})/r \tag{5.64}$$

and since γ^{SV} is virtually constant the important requirement is that γ^{SL} should be as small as possible. However, if θ is zero then

$$P = -2\gamma^{LV}/r \tag{5.65}$$

so that now γ^{LV} should be large. This suggests that to achieve maximum penetration one needs a low γ^{SL} with γ^{LV} as high as possible, and in practice these apparently opposing effects are optimised by careful choice of tenside.

In capillary systems of practical interest there is besides contact angle and interfacial tension, another important factor that determines the driving force flow, namely the geometry of the solid surface at the boundary line between the solid, liquid and third phase (vapour or second immiscible liquid). Schwartz [44] has described the more commonly encountered shapes of liquids in contact with solids as the shapes of revolution of unduloid, catenoid and nodoid. Calculations of capillary penetration based on simple geometry (as is equation (3.10)) are possible, but systems containing complex capillary structures are usually not amenable to this approach. Furthermore the relevant contact angle may not be readily defined — hysteresis is quite common and for textile fibres this may be as high as 50-60° with the receding angle close to zero. The advancing angle is normally used in capillarity calculations.

Equation (5.63) indicates that liquid will spontaneously penetrate the channels between textile fibres (or powder particles) only if $\theta < 90°$. Hence this establishes the requirement for making a fabric water repellent, and is achieved by coating the fibres with a material, for example silicones having a high contact angle with water. A fabric is made up of a regular array of fibres, and a geometric structure is not too difficult to describe, but this is certainly not the case with powders whose internal structure is of irregular dimensions determined by the size, shape and surface texture of the particles. For complete wetting of all the internal surfaces air must escape, and since a variety of possible configurations arise ranging from channels closed at one end to channels with different end dimensions, the exact description of the complete wetting process on a microscopic scale becomes impossible. Theoretical work in this area is limited, but suggests that zero contact angle is a necessary pre-requisite for the total displacement of air. A low contact angle is also required for wetting fabrics, and in both cases γ^{LV} to be as high as possible — again these two requirements are apparently contradictory, and in practice θ is kept low by the use of tensides.

Although completeness of air removal is important, the time required to achieve it is most relevant. The linear rate of laminar flow in a cylindrical tube is described by Poiseuille's formula

$$\frac{\mathrm{d}h}{\mathrm{d}t} = \frac{r^2 P}{8\eta h} \tag{5.66}$$

where η is the viscosity of the liquid, and h the length of liquid column at time t. For a horizontal capillary, or in general where gravity can be neglected,

$$h = (rt\gamma^{LV} \cos \theta / 2\eta)^{\frac{1}{2}} \tag{5.67}$$

or the length of the liquid column is proportional to the square root of the time that has elapsed after immersion. Equation (5.67) is usually ascribed to Washburn [45], and has been experimentally verified. It can be suitably modified for a packed bed of solid particles by replacing the radius by a factor K which contains an effective radius for the bed and a tortuosity factor that allows for the complex path taken by the liquid through the channels between the particles. Provided the packing of the powder bed remains constant during wetting, the linear relationship between h^2 and t is readily demonstrated [46]. For rapid penetration high $\gamma^{LV} \cos \theta$, low θ and low η are desirable, with K as large as possible, that is, a loosely packed powder. But as before, high γ^{LV} and low θ are not normally compatible, so in practice it is θ which is kept as small as possible.

5.4.3 Detergency
Detergency is the name given to the process of removal of foreign matter from solid surfaces by the application of surface chemistry, and the washing of clothes is a good example of detergent action. The 'soil' that accumulates on fabric contains a variety of materials including oils and greases and solid particles of dust, soot etc., and these are commonly removed by washing the fabric in an aqueous detergent solution.

A number of factors are involved in the total washing process. The simple requirement is that the fabric-soil interface is replaced by fabric-solution and soil-solution interfaces; the work involved in this process is given by

$$W = \gamma^{OL} + \gamma^{SL} - \gamma^{OS} \tag{5.68}$$

where O represents the soil. The condition for spontaneity is

$$\gamma^{OS} \geqslant \gamma^{OL} + \gamma^{SL} \tag{5.69}$$

so that the desired effect is in principle obtained by making γ^{OL} and γ^{SL} as small as possible without a major change in γ^{OS}. This could be achieved with a tenside that is strongly adsorbed from aqueous solution on both soil and fabric. Some mechanical action may be required to assist the separation, and therefore agitation is normally a part of the washing process.

For a fluid soil adhering to fibre, the contact angles are usually the controlling factor. Essentially the requirement is to increase the angle subtended by the oil on the fibre from 0 to 180°, and this occurs by adsorption of tenside at both fibre and oil surfaces with corresponding reduction in γ^{OL} and γ^{SL}. High temperature, to reduce viscosity, and mechanical agitation are usually involved since both assist the oil removal. The detergent action is illustrated in Fig. 5.10 which shows how oil rolls up into globules on wool fibres by the action of detergent.

Following displacement of the soil from the fabric it is necessary to prevent redeposition, so the detergent must have 'suspending' power, that is, the soil remains dispersed in the fluid phase as a dispersion or emulsion (for oils) and is removed with it at the end of the process. These and other aspects of detergent action are discussed in Durham's book [47].

5.4.4 Flotation
One of the most complex applications of surface chemical principles is the flotation process in which various types of solid particles are separated from each other; froth flotation is the most widely used procedure and this involves suspending the solid mixture in a solution of a foaming agent through which air is bubbled. For example a powdered mixture of zinc sulphide and felspar minerals are readily separated by froth flotation, since the former becomes attached to the air bubbles and are carried upward into the foam, while the felspar remains in the liquid phase. The particles are usually very much smaller than the air bubbles, so many particles may be attached to each bubble. Special additives (collectors) having a hydrocarbon chain and a strongly adsorbing head group, are employed to bring about the required increase in contact angle and improve the efficiency of separation.

Essentially the particles to be floated are transferred from the liquid phase, in which they are completely wetted, to a stable position at the liquid-air (bubble) interface, which is then carried to the surface of the liquid. It is readily shown [48] that particles forming a finite contact angle at the air-liquid interface are stable at the interface, and for efficient separation the angle should be as large as possible while that of the other component is kept close to zero. The collectors selectively adsorb at the particle surface and cause the desired interfacial energy changes. Optimum flotation usually occurs as monolayer coverage of collector is approached, for example silver iodide floats optimally when the surface is covered with cationic tenside adsorbed in parallel orientation [49]. Frothing agents are added, which are designed to ensure stability of the bubbles but also contribute to the interfacial energies and as such help to stabilise the particle attachment to the bubble. For further detail the reader is referred to Gaudin's book [50].

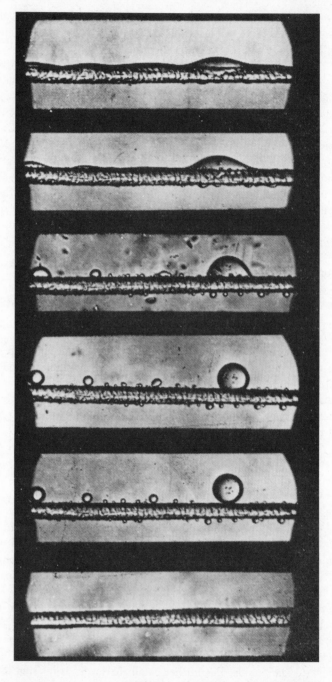

Fig. 5.10 – Stages in rolling-up of oil on a wool fibre after addition of detergent.

5.5 REFERENCES

[1]   A. I. Bailey and S. M. Kay, (1967) *Proc. Roy. Soc.* **A301**, 47.
[2]   F. M. Fowkes, (1967) *Wetting,* Soc. Chem. Ind. London Monograph, **25**, 3.
[3]   F. M. Fowkes, (1964) *Ind. Eng. Chem.,* **56**, 40.
[4]   W. A. Zisman, (1964) *Adv. in Chem.,* **43**, 1.
[5]   R. E. Johnson and R. H. Dettre, (1969) *Surface and Colloid Science,* Ed. E. Matijevic, Wiley-Interscience, New York, Vol 2, p. 85.
      T. D. Blake and J. M. Haynes, (1973) *Progress in Surface and Membrane Science,* Eds. J. F. Danielli, M. D. Rosenberg and D. A. Cadenhead, Academic Press, New York, Vol 6, p. 125.
[6]   J. F. Padday, (1968) *J. Colloid Interface Sci.,* **28**, 557.
[7]   T. D. Blake, J. L. Cayias, W. H. Wade and J. A. Zerdecki, (1971) *J. Colloid Interface Sci.,* **37**, 678.
[8]   B. V. Derjaguin, (1965) *XXth I.U.P.A.C. Congress,* Butterworths, London, p. 375.
[9]   A. C. Zettlemoyer and K. S. Narayan, (1967) *The Solid-Gas Interface,* Ed. A. E. Flood, Edward Arnold Ltd., London, Vol.1, Ch. 6.
[10] A. C. Zettlemoyer, (1969) *Hydrophobic Surfaces,* Ed. F. M. Fowkes, Academic Press, New York, p. 1.
[11] J. H. Clint, (1973) *J. Chem. Soc. Faraday Trans., 1,* **69**, 1320; D. H. Everett and G. H. Findenegg, (1969) *J. Chem. Thermodynamics, 1,* 573.
[12] G. H. Findenegg, (1972) *J. Chem. Soc. Faraday Trans., I,* **68**, 1799.
[13] G. H. Findenegg, (1971) *J. Colloid Interface Sci.,* **35**, 249.
[14] G. H. Findenegg, (1973) *J. Chem. Soc. Faraday Trans. I,* **69**, 1069.
[15] W. Drost-Hansen, (1969) *Ind. Eng. Chem.,* **61**, 10.
[16] G. Peschel and P. Belouschek, (1979) *Cell-Associated Water,* Ed. W. Drost-Hansen and J. S. Clegg, Academic Press, New York, p. 3.
[17] F. H. Healy, J. J. Chessick, A. C. Zettlemoyer and G. J. Young, (1954) *J. Phys. Chem.,* **58**, 887.
[18] J. J. Chessick, A. C. Zettlemoyer, F. H. Healey and G. J. Young, (1955) *Can. J. Chem.,* **33**, 251.
[19] J. A. Lavelle and A. C. Zettlemoyer, (1967) *J. Phys. Chem.,* **71**, 414.
[20] W. Ostwald and R. de Izaquirre, (1922) *Kolloid Zeits.,* **30**, 279.
[21] G. Schay, (1969) *Surface and Colloid Science,* Ed. E. Matijevic, Wiley-Interscience, New York, Vol. 2, p. 155.
[22] R. E. Day, Yu. A. Eltekov, G. D. Parfitt and P. C. Thompson, (1969) *Trans. Faraday Soc.,* **65**, 266.
[23] D. H. Everett, (1964) *Trans. Faraday Soc.,* **60**, 1803.
[24] D. H. Everett, (1965) *Trans. Faraday Soc.,* **61**, 2478.
[25] J. J. Kipling, (1965) *Adsorption from Solutions of Non-Electrolytes,* Academic Press, London.
[26] R. E. Day and G. D. Parfitt, (1967) *J. Phys. Chem.,* **71**, 3073.

[27] S. G. Ash, D. H. Everett and G. H. Findenegg, (1968) *Trans. Faraday Soc.*, **64**, 2645.

[28] D. H. Everett, (1978) *Progr. Colloid Polymer Sci.*, **65**, 103.

[29] A. W. Adamson, (1976) *Physical Chemistry of Surfaces*, 3rd ed. Wiley, New York, p. 389.

[30] R. R. Stromberg, L. E. Smith and F. L. McCrackin, (1970) *Symposia Faraday Soc.*, **4**, 192.

[31] E. Killmann, J. Eisenlauer and M. Korn, (1977) *J. Polymer Sci., Polymer Symposium*, **61**, 413.

[32] E. J. Clayfield and E. C. Lumb, (1966) *J. Colloid Interface Sci.*, **22**, 285.

[33] M. Lal and R. F. T. Stepto, (1977) *J. Polymer Sci., Polymer Symposium*, **61**, 401.

[34] A. Silberberg, (1971) *Pure and Applied Chem.*, **26**, 583.

[35] K. J. Mysels, (1959) *Introduction to Colloid Chemistry*, Interscience, New York, Ch. 15.

[36] A. L. Smith, (1973) *Dispersion of Powders in Liquids*, 2nd ed., Ed. G. D. Parfitt, Applied Science Publishers, London, Ch. 3.

[37] A. W. Adamson, (1976) *Physical Chemistry of Surfaces*, 3rd ed., Wiley, New York, Ch. IV.

[38] C. H. Giles, T. H. MacEwan, S. N. Nakhwa and D. Smith, (1960) *J. Chem. Soc.*, 3973.

[39] F. G. Greenwood, G. D. Parfitt, N. H. Picton and D. G. Wharton, (1968) *Adv. in Chemistry Series*, **79**, 135.

[40] D. W. Fuerstenau, (1971) *The Chemistry of Biosurfaces*, Ed. M. L. Hair, Marcel Dekker Inc., New York, Vol. 1, Ch. 4.

[41] G. J. Young, J. J. Chessick and F. H. Healey, (1956) *J. Phys. Chem.*, **60**, 394.

[42] D. H. Everett, D. F. Billett and E. H. M. Wright, (1964) *Proc. Chem. Soc.*, 216.

[43] G. D. Parfitt and M. W. Tideswell, (1980) *J. Colloid Interface Sci.*, in press.

[44] A. M. Schwartz, (1969) *Ind. Eng. Chem.*, **61**, 10.

[45] E. D. Washburn, (1921) *Phys. Rev.*, **17**, 374.

[46] V. T. Crowl and W. D. S. Wooldridge, (1967) *Wetting*, Soc. Chem. Ind. London Monograph, 25, 200.

[47] K. Durham, Ed. (1961) *Surface Activity and Detergency*, Macmillan, London.

[48] A. W. Adamson, (1976) *Physical Chemistry of Surfaces*, 3rd ed., Wiley, New York, p. 480.

[49] M. J. Jaycock and R. H. Ottewill, (1963) *Bull. Inst. Mining Met.* No. 677, 497.

[50] A. M. Gaudin, (1957) *Flotation*, McGraw-Hill, New York.

"Alice waited a minute to see if he would speak again, but, as he never opened his eyes or took any further notice of her, she said 'Good-bye!' once more, and, getting no answer to this, she quietly walked away: but she couldn't help saying to herself, as she went, 'Of all the unsatisfactory —' (she repeated this aloud, as it was a great comfort to have a such a long word to say) 'of all the unsatisfactory people I *ever* met —' She never finished the sentence, for at this moment a heavy crash shook the forest from end to end."

Through the Looking Glass
and What Alice Found There

Index